# BEYOND BALI

## STRATEGIC ISSUES
## FOR THE
## POST-2012 CLIMATE CHANGE REGIME

EDITED BY
CHRISTIAN EGENHOFER

WITH A PREFACE BY
BO KJELLÉN & FRANK CONVERY

PRINCIPAL AUTHORS
Markus Åhman, Thomas L. Brewer,
Christian Egenhofer, Carolyn Fischer,
Ned Helme, Niklas Höhne,
Lars G. Josefsson, Sivan Kartha,
Adrian Macey, Surya P. Sethi,
Anders Wijkman, Lars Zetterberg

OTHER CONTRIBUTORS
Monica Alessi, Frank Convery,
Erik Haites, Bo Kjellén, Diana Movius,
Matt Ogonowski, Jake Schmidt,
Farhana Yamin

The European Climate Platform (ECP) is a joint initiative of the Centre for European Policy Studies (CEPS) in Brussels and the Climate Policy Research Programme (Clipore) of the Swedish Foundation for Strategic Environmental Research (Mistra) in Stockholm. Established in 2005, the ECP aims to facilitate interaction within the policy research community, mainly but not exclusively in Europe. Its working methods consist of bringing together a select number of policy-makers, negotiators and experts to vigorously debate key topics in the area of international climate change policy and to widely disseminate its conclusions. The ECP actively seeks dialogue with policy-makers and other stakeholders while being dedicated to academic excellence, unqualified independence and policy relevance. The ECP is governed by a steering group, drawn from government and academia.

ISBN 978-92-9079-738-8



Centre for European Policy Studies
Place du Congrès 1, B-1000 Brussels
Tel: 32 (0) 2 229.39.11 Fax: 32 (0) 2 219.41.51
e-mail: info@ceps.eu
internet: http://www.ceps.eu

# CONTENTS

# PREFACE

The European Climate Platform (ECP) has been established as a partnership between CEPS and Clipore, the Climate Policy Research Programme of the Swedish Foundation for Strategic Environmental Research (Mistra). It brings together policy practitioners and members of the research community to explore how the climate change agenda can be productively advanced. It operates at two levels: the first is the preparation of empirically-based analyses by leading researchers in the field; the second is informal but structured debate and interaction on these analyses in Brussels by policy practitioners and analysts. These meetings allow policy practitioners - and especially negotiators - to step back and engage in discussions on the wider strategic issues that can help shape their subsequent deliberations and negotiations. They enable policy analysts and the research community to understand the realities that must be addressed before progress is to be achieved.

It is difficult to assess precisely how productive this interface between the academic and the practical is, but we know, as joint chairs of the seminars, that they have been informed, enjoyable and helpful to those in the middle of the policy process to better understand the evidence and the wider context, and to the research community to know what are the key issues and priorities for research.

It has been a pleasure and a privilege for us to play our small part in keeping this important and productive dialogue fresh and useful. The mutual enrichment that comes from interaction is perhaps the most important product of this effort. But it is also important to project insights and outputs to the wider community that was not fortunate enough to be party to these seminars. And this is what this volume aspires to be - a symbol and a manifestation of our endeavours. We expect that, in the swirling cacophony of paper, images and words that surrounds the Bali process and its aftermath, the insights this volume offers will strike a chord and make a contribution.

This book is a direct outcome of the ECP High-Level Colloquium on Climate Change: Key Actions for the Crucial Years ahead, which was held in Brussels on 22-23 March 2007. This event brought together more than 100 climate change negotiators from all other the world, government and EU officials, researchers and business and NGO representatives.

We wish to express our gratitude to Deborah Cornland for her contribution in launching the ECP, and to thank her for her support. Deborah served as the first Programme Director of Clipore until she stepped down from this position in 2007. Without her enthusiastic leadership, this book as well as the ECP would not have been possible.

Frank Convery
Co-Chairman, ECP &
Heritage Trust Professor
of Environmental Policy,
University College Dublin

Bo Kjellén
Co-Chairman, ECP &
Former Chief Climate Change
Negotiator of Sweden

# 1. FROM INCONVENIENT TRUTHS TO A COPENHAGEN PROTOCOL?

*CHRISTIAN EGENHOFER*[*]

Only the biggest optimist would have predicted even one year ago that governments participating in the global climate negotiations at Bali, Indonesia (COP13/CMP3) would agree on a roadmap towards a global climate change agreement to be completed by the end of 2009, ready to fill the gap when the commitments under the Kyoto Protocol expire in 2012. Sure, agreement on the Bali Action Plan was only reached after many days of exhausting negotiations, which at times hovered on the brink of collapse. The primary issues were whether or not indicative percentage reduction targets of 25-40% should be included in the mandate and the precise wording to be to used to allow discussion of reduction commitments for developing countries. Both issues lie at the heart of the climate change negotiations and will undoubtedly continue to prove extremely controversial in the future. But in essence, these are the questions that ministers and negotiators will have to resolve in the final hours of Copenhagen in 2009.

What is far more important, however, is that the Bali Action Plan firmly sets out the four key elements that should be considered – mitigation, adaptation, technology and financing – by the new Ad Hoc Working Group on Long-term Cooperative Action under the Convention, with a mandate to complete its work in 2009. All governments supported an end date of 2009.

---

[*] Christian Egenhofer is a Senior Research Fellow at both CEPS and the Centre for Energy, Petroleum and Mineral Law and Policy at the University of Dundee, Scotland and serves as a member of the ECP Steering Group.

There have been additional signs that the international community is seriously willing this time round to strike a deal. For the first time, finance ministers met during the climate change negotiations to better understand the scale of the challenge and to explore the potential need for transfers, which undoubtedly will be required in the final hours of the negotiations. Almost in parallel, there has been a first meeting among trade ministers on climate change issues in the so-called 'Informal Trade Ministers Dialogue on Climate Change' (ITMD), initiated by the government of Indonesia. Essentially, it was an exchange of views leading to a number of procedural recommendations calling inter alia for an urgent and successful conclusion of the Doha Development Agenda negotiations, and for more comprehensive research and better empirical evidence on the linkages and interface between international trade developments. This reinforces the impression that the momentum towards a global deal is gathering pace.

Such momentum will be needed. The Intergovernmental Panel on Climate Change (IPCC) estimates that the world will have to find about 30 gigatonnes of $CO_2$ equivalent reductions from baseline by 2030 to reach the lower stabilisation levels of 440-490 parts per million of $CO_2$ equivalent. This means globally at least 50% below 2000 levels by 2050 or for developed countries listed in Annex I, 25-40% below 1990 levels by 2020.

The contributions in this book examine some of the most difficult and controversial questions that global climate change negotiators face between now and the emergence of a 'Copenhagen Protocol' in 2009, and even beyond. The different chapters to explore the four principal elements in the Bali Action Plan – mitigation, adaptation, technology and financing from different country, stakeholder or political perspectives.

Fortunately, there is a long history of negotiations and a stock of already-acquired knowledge on a future climate change regime upon which the Bali Action Plan can rely. They are presented in chapter 2. The different discussion processes both within the UNFCCC but also outside, such as the G8+5, the Asia Pacific Partnership on Clean Development and Climate and the Major Economies Meetings, seem to lead to a more enhanced consideration of the issue of climate change at the international level. On the other hand, the urgency of action that is suggested by climate science has yet to lead to accelerated analytical work that could feed into negotiations on the time scale needed to ensure that there is no gap between the first and second commitment periods. And even if such acceleration were to happen, it would equally require that negotiators are capable of absorbing the increasing body of analytical work. There is no guarantee that the various future action-related processes and the Bali Action Plan can deliver a result in a timeframe that is able to avoid dangerous interference with the climate system. This is where political will be required, a recurrent theme throughout the book.

Chapter 3 makes the case that estimating the global mitigation potential will need to be at the heart of any effort towards the daunting task of reducing reduce greenhouse gas emissions to safe levels. If the mitigation potential is properly identified, attention can turn to how it is realised. However, translating the potential into action will require better organisation, more resources and a firm international commitment. The review of ongoing activities to estimate mitigation potential in different fora or reports such as the Fourth Assessment Report, the UNFCCC, the Stern Review and especially the Ad-hoc Working Group on Future Commitments for Annex I Parties under Article 3.9 of the Kyoto Protocol (AWG) document the impressive progress that has been made in this area. The review however also reveals the many methodological difficulties the concept of 'mitigation potential' faces; for example, mitigation potential can be viewed from several perspectives, including market potential, economic potential, socio-economic potential, technological potential and physical potential.

One conclusion of chapter 3 is that after 2012, there is likely to be a much larger spread around an average figure for GHG reduction commitments than in the first commitment period. At the same time, mitigation potentials will evolve over time, as the concept is a dynamic one. Evolution in the type of economic activity - for example, moving from a manufacturing to a service economy, or technological breakthroughs - could increase the mitigation potential. If the potential were to increase, countries could take on more ambitious targets. This argues for post-2012 arrangements having built-in flexibility, unlike the first commitment period where no change was possible even though the commitments were negotiated over ten years before they became effective. Various possible instruments for flexibility exist - regular review, breaking down targets by sectors, sectoral agreements, some form of adjustment mechanism according to changes in economic or technological factors. The flexibility, however, would find its limits in the necessity for stability of carbon markets. One suggestion is to make post-2012 targets provisional before a fully-fledged, legally binding global agreement is in place. This could give the market a positive signal, without reducing the necessary flexibility pending final decisions.

Chapter 4 provides a sobering perspective from a fast-growing developing country where GHG mitigation policies compete with other priorities, notably development and adaptation. Developing countries are faced with numerous 'inconvenient truths':

- Climate change is already happening because of natural drivers duly accelerated by anthropogenic activity.
- Industrialised countries have consumed the bulk of the global carbon budget while developing in an unconstrained world, thereby threatening to leave limited headroom for developing countries.
- A climate-constrained world might actually impose real limits to growth; if so, distributional conflicts will accelerate.
- At the same time, the number of poor will continue to grow, representing roughly one-third of total world population. People dependent on biomass to meet up to 90% of their household energy needs will even rise from 2.5 billion today to 2.7 billion by 2030. The world will need to accept a threshold level of development for the poor and to make this a priority over all other global endeavours.
- It is well known that the emissions from industrialised countries must peak soon, calling current lifestyles and patterns of production and consumption into question. It is clear that addressing this will entail huge costs and it is near impossible to see how that would be politically impossible without a binding global agreement with targets.
- And still industrialised countries would need to fund even the no-regret mitigation options in developing countries since developmental outlays will take precedence over mitigation in a resource-constrained developing country.
- Even if industrialised countries succeed in bringing down their emissions to a level that is 80% below their 1990 level by 2050, their emissions would still be a multiple of its fair share under a per capita metric.
- As technological breakthroughs by 2030 are very unlikely, the bet is on making current low-carbon and energy-efficient technologies universally available and promote collaborative research.

Neither developed nor developing countries on their own can effectively address global climate concerns. Cooperation and collaboration on a much broader scope and scale is likely to be needed. For example, we should expect a post-Kyoto regime with more collaborative efforts in the future energy research. This might entail appropriate sharing of Intellectual Property Rights (IPRs) to allow that available energy efficient and climate friendly technologies are put into limited public domain to avoid a carbon-intensive and business-as-usual developing country growth trajectory that repeats the development paths of industrialised countries. While this is a difficult and delicate issue, discussion on it cannot be avoided. The post-Kyoto regime will also need to recognise that energy security is a global need and not just the right of some countries. From a developing countries perspective such as India, a post-Kyoto regime would seek to curb unsustainable lifestyles wherever they exist, renew commitment to meeting the Millennium Development Goals and eradicating poverty; the additional funding to achieve this would need to come from developed countries, although in different possible forms such as carbon markets, global funds or direct assistance by developed countries.

Chapter 5 zooms in on climate change as threat to development, a regularly neglected issue. The topic of climate change remains remote to people living in low-income countries, compared to more immediate problems like poverty, disease and hunger, although the consequences of climate change will affect development in a major way. However, climate change will significantly increase the frequency and strength of extreme weather events. Many regions of the world already have a long history of serious disaster-related problems. But natural disasters occur only when communities are exposed to potentially hazardous events without being able to absorb the impact. While it is common to talk about natural disasters, it is too often forgotten that both vulnerability and hazard are conditioned by human activities. The international response to millions of poor people who are already experiencing the effects of climate change is insufficient. It is commendable that the Swedish Government recently launched an International Commission on Risk Reduction and Adaptation to explore and promote effective ways to integrate risk reduction and adaptation to climate change into development and poverty reduction plans in developing countries and to secure future investment in development aid to take full account of climate stresses and increased disaster risks. In a practical sense, this means: i) to identify and analyse the incentives as well as the barriers for developing countries to undertake risk-reduction and climate-proofing measures in their development efforts; ii) to consider how to best combine long-term efforts for climate change mitigation and the urgent need to support adaptation effort; iii) to map out directions for international development cooperation in the field of adaptation and risk reduction; and iv) to consider how to achieve policy coherence by integrating concerns for climate change in wider development efforts.

Chapter 6 discusses how climate change affects business. It draws a useful distinction between the ideal situation with adequate public and private responses and the actual situation, where industry faces increased risks. Unless international and national policy-makers respond adequately, uncertainty in business conditions will continue to increase. Whenever there is uncertainty, for example if a sector is threatened by regulation, investments are put on hold until the uncertainty has disappeared. Current global climate policy exemplifies this uncertainty, which spreads in many ways. Managers are faced daily with the questions of which sectors will be included in climate policy, whether global transport will continue to be exempt from GHG policies, and when will Europe and other signatories to the Kyoto Protocol be joined in their efforts. If emissions from production plants can be 'exported' from Europe only to be 'imported' again in the form of finished products, a politically-led, relatively short-term allocation becomes a determining factor in long-term investment decisions. Politicians must not exacerbate this uncertainty by taking short-term, ambiguous decisions. The clearer the rules of the game are, the easier will be the transformation to a low-emission economy.

The challenge of adequately addressing adaptation to climate change impacts in developing counties by means of international collaboration is examined in chapter 7. It finds that adaptation work has matured from discrete, localised solutions towards a more comprehensive development-based perspective. This has drawn policy-makers' attention towards the possibility of making adaptation part of development policy, which is referred to as 'mainstreaming'. Neverhteless, developed countries' development agencies still appear to attach insufficient attention to adaptation. Yet the same can be said of developing countries and their policy priorities.

One of the thorny issues is funding. Given the magnitude of the challenge, adequate adaptation will require considerable additional financial flows from developed countries to developing countries. The key to unlocking significant financial flows seems to lie with a reasonable assurance that developing countries have enough absorptive capacity to use those funds towards effective and efficient adaptation activities. This will require improvements in knowledge and expertise, institutional strength, good governance, transparency and accountability. The chapter argues that the discussion about the source of the financing (i.e. how to share the burden among developed countries) should not wait until all questions on implementation (i.e. how the financial resources are used for developing country adaptation) are resolved. A too narrow link between the two might leave us in a chicken-and-egg situation. The chapter finishes with a number of concrete and operational recommendations. It sets out a sequence of steps towards the evolution of a global adaptation regime, defines the linkages between adaptation and human development, before ending with financing and the need for further informal dialogues.

The two following chapters deal with technology development and diffusion. Chapter 8 takes stock of existing activities on technology and innovation for both developed and developing countries, including finance and investment issues. The chapter reviews all major approaches within the UNFCCC (e.g. the Expert Group on Technology Transfer, the Global Environmental Facility and the Clean Development Mechanism) or outside (such as the Climate Technology Initiative under the Gleneagles Plan of Action, the Asia Pacific Partnership on Clean Development and Climate, the Major Economies Meetings or sectoral approaches) and groups them according to four categories of technology-oriented agreements (TOAs):

1) knowledge-sharing and coordination,
2) research, development and demonstration (RD&D),
3) technology transfer and
4) technology deployment mandates, standards, and incentives.

It finds that to date, most existing TOAs fall into the first category, i.e. knowledge-sharing agreements that are relatively low cost by focusing on information exchange, the promotion of common standards or horizontal innovation schemes. Their effectiveness is limited by the voluntary, non-binding nature of the frameworks, however. Technology cooperation agreements, i.e. category 2, which primarily focuses on fundamental research and demonstration projects exist, but are few in number. Most likely such agreements will find their limits, however, when the technology moves from the pre-competitive to the post-competitive stage. Technology transfer, i.e. category 3, has been attempted within the UNFCCC. However, simply transferring patents is not enough, as most technology transfer depends on know-how. Category 4, technology deployment mandates offer particular potential in the form of harmonised standards for renewable energy, building codes, energy efficiency or requirements for carbon capture and sequestration. Yet, there are few, if any successful examples of this kind of agreement. All of these categories are useful, however, and should simultaneously be expanded.

The complementary technology chapter 9 by Tom Brewer proposes a taxonomy for structuring analyses and negotiations. Thereby it focuses on the needs of analysts and negotiators who are involved in negotiations for a post-2012 climate regime. In concrete terms, the taxonomy is in four parts, each of which is focused on a particular element of a decision-making process: i) describing technologies (What are their key features? How are they similar and different?); ii) identifying the context of technologies (How do economic and political factors affect the development and use of technologies?); iii) evaluating technologies (What goals and criteria are relevant for evaluating technologies?) and iv) pursuing pathways into the future (How can the development and use of mitigating or adaptive technologies be facilitated? How can detrimental technologies be constrained?). These elements are then discussed separately in the chapter and expressed in the form of an analytic tree in the concluding section.

Chapter 10 focuses on encouraging actions to reduce GHG emissions from deforestation and degradation in developing countries (REDD) through the so-called 'dual markets approach'. This approach combines elements of market and fund-based approaches. With approximately 20% of carbon dioxide emissions globally coming from tropical deforestation, REDD is crucial to addressing global climate change.

By creating a separate REDD market, in contrast to other market-based proposals, the Dual Markets framework would maintain the integrity of the existing carbon market, while generating more financing than an exclusively fund-based approach. The idea is that developed (i.e. Annex I) countries agree to specific REDD targets, providing a guaranteed level of demand for developing country REDD credits while safeguarding stabilisation goals against potential uncertainties of REDD reductions. The proposal would provide a range of options to both developing and developed countries, including voluntary participation, borrowing from future REDD targets, making up REDD shortfalls with carbon market credits and review by the COP to consider linking to the post-2012 carbon market. The dual markets approach would facilitate a system of learning-by-doing for the upcoming post-2012 commitment period and allow the international community to craft a workable incentive system to reduce emissions from deforestation. Such a REDD mechanism would be capable of addressing many issues, including the establishment of reference scenarios and baselines, creating effective incentives for developing and developed countries and ensuring a sufficient and consistent flow of funding. Fortunately, Parties have submitted a variety of proposals for an international REDD mechanism, relying on carbon market integration and/or voluntary contributions.

The two concluding chapters 11 and 12 deal with international aviation and maritime emissions, a small, but rapidly increasing, percentage of global greenhouse gas emissions. A business-as-usual scenario could put emissions from international aviation and maritime shipping or so-called 'bunker fuels' at 15% of global emissions by 2050. To date, very little has been done to curb this trend. This development, coupled with the growing political popularity of emissions trading, has put the spotlight on the option of an integrated international trading system, including international aviation and shipping. The idea would be to address emissions from international aviation and shipping and capture the cost savings that a broad emissions trading system offers.

However, the characteristics of bunker fuel emissions, with close links to international trade and competition, the difficulties of allocating emissions on a country level and the complexities in estimating climate impact from emissions, make the feasibility and effectiveness of a conventional emissions trading system questionable. Furthermore, a full integration of international transport and industry could cause significant negative effects for industry, without achieving substantial cuts in bunker-fuel emissions. In the medium term at least, these negative effects seem to offset the advantages of an integrated system.

The two chapters come to different conclusions from this analysis. Chapter 11 considers a sectoral approach - although not yet defined in detail - as a more promising option for bunker fuels than incorporating them into an integrated emissions trading system. Even though this would, in theory, raise the total cost of reaching a global emissions target as compared to an integrated emissions trading system, it offers several advantages. Most importantly, it seems feasible from a technical and political point of view and would allow the negative effects on industry to be managed, without compromising an effective control of emissions from international aviation and shipping.

Chapter 12 proposes a separate emissions trading system (ETS) for international transport alone in which transport emissions could effectively be controlled. The impact on the industrial allowance price can be reduced, at least to some extent. But even with separate systems, the transport sector will indirectly influence the allowance price in industry. Both sectors will compete for the same emissions reductions such as bio energy, clean electricity and CERs (certified emissions reductions) or other offsets. The transport sector is likely to increase the demand for these solutions and thus indirectly increase the allowance price in the industrial ETS.

These are only some of the thoughts and proposals that are elaborated in greater depth in the following chapters. By addressing the difficult and controversial topics that global climate change negotiators face, the chapters make a contribution to the task of finding a new global framework for avoiding dangerous climate change beyond 2012.

# Part I
# Political Priorities

# 2. THE HISTORY AND STATUS OF THE INTERNATIONAL NEGOTIATIONS ON A FUTURE CLIMATE AGREEMENT

*NIKLAS HÖHNE, FARHANA YAMIN &ERIK HAITES*[*]

## 1. Introduction

The objective of the United Nations Framework Convention on Climate Change (UNFCCC) is to stabilise greenhouse gas (GHG) concentrations to avoid dangerous anthropogenic interference with the climate system. Several countries, including the 27 member states of the European Union, and many environmental NGOs have agreed that the global average temperature increase should be limited to 2°C above pre-industrial levels to avoid such dangerous interference. To have a 50% probability of achieving this target, atmospheric concentrations of greenhouse gases would need to be stabilised at no more than 450 ppmv $CO_2$eq (parts per million by volume of carbon dioxide equivalent) (Meinshausen, 2005).

To stabilise atmospheric concentrations, current global emissions of greenhouse gases must be reduced significantly. The lower the desired concentration, the sooner the decline in global emissions must start. Under the UNFCCC and the Kyoto Protocol, participating countries agreed on different commitments and emissions reduction targets. To start to reduce global emissions, more substantial reductions of greenhouse gases are necessary.

[*] Niklas Höhne is a consultant at Ecofys in Cologne; Farhana Yamin is a researcher at the Institute of Development Studies, University of Sussex and Erik Haites is with Margaree Consultants in Toronto.

This chapter gives an overview of the history of the negotiations under the climate change regime and the resulting commitments under the UNFCCC and the Kyoto Protocol. It also summarises current activities relating to the negotiation of an agreement setting out commitments over and beyond those made in the UNFCCC/Kyoto Protocol and the arguments advanced over whether such an agreement should be within or outside the UNFCCC. Finally, the paper gives more detailed information on the negotiating positions of important countries or groups on the timing, scope and nature of such negotiations.

## 2. Current commitments under the UNFCCC and the Kyoto protocol

This section provides an overview of the current commitments of countries under the UNFCCC regime. It is based on Depledge (2000), Oberthür & Ott (1999), Yamin & Depledge (2004) and Höhne (2006).

The UNFCCC's ultimate objective is to stabilise greenhouse gas concentrations at a level that would prevent dangerous anthropogenic interference with the climate system (see Box 1).

*Box 1. Article 2 of the UNFCCC*

The ultimate objective of this Convention and any related legal instruments that the Conference of the Parties may adopt is to achieve, in accordance with the relevant provisions of the Convention, stabilization of greenhouse gas concentrations in the atmosphere at a level that would prevent dangerous anthropogenic interference with the climate system. Such a level should be achieved within a time-frame sufficient to allow ecosystems to adapt naturally to climate change, to ensure that food production is not threatened and to enable economic development to proceed in a sustainable manner.

To reach this goal, the UNFCCC builds upon the principle of common but differentiated responsibilities and capabilities of Parties. Accordingly, countries are divided into three groups for the purpose of differentiating the obligations or commitments under the Conventions and later the Kyoto Protocol (see also Table 1):

a) *Parties included in Annex II to the Convention:* countries that were members of the Organisation for Economic Cooperation and Development (OECD) in 1992.

*b)* *Parties included in Annex I to the Convention (Annex I Parties):* both the Annex II Parties and countries with 'economies in transition' (EITs), the Russian Federation and several other Central and Eastern European countries.

c) *Parties not included in Annex I to the Convention (non-Annex I Parties):* those countries that are not listed in Annex I, including all newly industrialised countries and developing countries.

*Table 1. Members of Annex I and their commitment under the Kyoto Protocol (Annex B)*

| **Country** | **Member of Annex I** | **Member of Annex II** | **Economies in transition** | **Commitment inscribed in Annex B (within parenthesis the outcome of the EU burden-sharing agreement)** |
|---|---|---|---|---|
| Australia | X | X | | 108 |
| Austria | X | X | | 92 (87) |
| Belarus | X | | X | 92 [d] |
| Belgium | X | X | | 92 (92.5) |
| Bulgaria | X | | X | 92 |
| Canada | X | X | | 94 |
| Croatia | X [a] | | X | 95 |
| Czech Republic | X [a] | | X | 92 |
| Denmark | X | X | | 92 (79) |
| Estonia | X | | X | 92 |
| European Community | X | X | | 92 |
| Finland | X | X | | 92 (100) |
| France | X | X | | 92 (100) |
| Germany | X | X | | 92 (79) |
| Greece | X | X | | 92 (125) |
| Hungary | X | | X | 94 |
| Iceland | X | X | | 110 |
| Ireland | X | X | | 92 (113) |
| Italy | X | X | | 92 (93.5) |
| Japan | X | X | | 94 |

| | | | | |
|---|---|---|---|---|
| Kazakhstan | X [b] | | X | To be negotiated |
| Latvia | X | | X | 92 |
| Liechtenstein | X[a] | | | 92 |
| Lithuania | X | | X | 92 |
| Luxembourg | X | X | | 92 (72) |
| Monaco | X [a] | | | 92 |
| The Netherlands | X | X | | 92 (94) |
| New Zealand | X | X | | 100 |
| Norway | X | X | | 101 |
| Poland | X | | X | 94 |
| Portugal | X | X | | 92 (127) |
| Romania | X | | X | 92 |
| Russian Federation | X | | X | 100 |
| Slovakia | X [a] | | X | 92 |
| Slovenia | X [a] | | X | 92 |
| Spain | X | X | | 92 (115) |
| Sweden | X | X | | 92 (104) |
| Switzerland | X | X | | 92 |
| Turkey | X | [c] | | [e] |
| Ukraine | X | | X | 100 |
| United Kingdom | X | X | | 92 (87.5) |
| United States | X | X | | 93 |

[a] Added to Annex I at the third Conference of the Parties in Kyoto 1997 (COP 3).

[b] Added at COP 7 only for the purpose of the Kyoto Protocol (see FCCC/CP/2001/13/Add.4, section V.C).

[c] Deleted from Annex II by decision 26/CP.7.

[d] Belarus had not ratified the Convention when the Kyoto Protocol was adopted. A commitment was agreed as an amendment to the Protocol at COP/MOP 2 by decision 10/CMP.2. Belarus has to maintain a reserve of an additional 7% at the end of the first commitment period. The amendment and the target will only enter into force once ratified by at least three-fourths of the Parties to the Protocol.

[e] No limit specified. Turkey had not ratified the Convention when the Kyoto Protocol was adopted.

Under the Convention, *all* Parties have certain general commitments (Article 4.1, UNFCCC):

- to prepare national inventories of greenhouse gas emissions;
- to implement measures to mitigate climate change;
- to promote and cooperate in the development, application and diffusion, including transfer, of technologies, practices and processes that reduce greenhouse gas emissions;
- to preserve sinks and reservoirs of greenhouse gases;
- to cooperate in preparing for adaptation to the impacts of climate change;
- to promote and cooperate in research on climate change;
- to exchange information related to climate change;
- to promote and cooperate in education, training and public awareness related to climate change ; and
- to report information related to the above in 'national communications'.

*Figure 1. Country groupings*

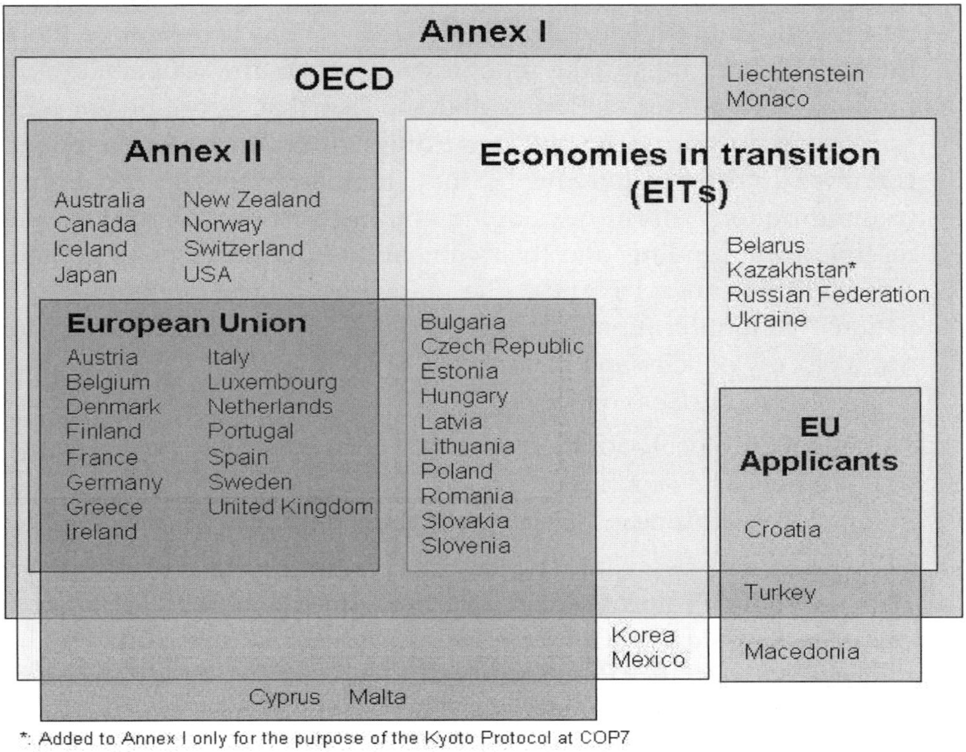

In addition to those general commitments, certain groups of countries have additional obligations or rights under the UNFCCC:

- Annex I Parties are to take the lead in modifying longer-term trends in emissions by adopting national policies and measures with the (not legally binding) original objective of returning their greenhouse gas emissions individually or jointly to 1990 levels by the year 2000 (Article 4.2, UNFCCC).
- The Parties included in Annex II have the further commitment to provide new and additional financial resources to meet the agreed full costs incurred by developing country Parties in complying with their obligations (Articles 4.3, 4.4, 4.5, UNFCCC).
- Economies in transition are allowed a certain degree of flexibility in implementing their commitments (Article 4.6, UNFCCC); for example, several of those countries have chosen a base year other than 1990.
- A Developing Country Party (a term that the COP has historically been unable to define so that it now simply means any Party not included in Annex I) is eligible for funding of the implementation of its general commitments (Article 11, UNFCCC). The extent of their implementation must take into account both the availability of funding from Annex II Parties and the fact that development and poverty reduction are the overriding priorities of developing countries. The requirement for regular reports ('national communications') from developing countries is also conditional on receipt of full funding, and their content and frequency are less than required for Annex I Parties. The guidelines for the preparation of national communications for non-Annex I Parties, for example, do not speak of "policies and measures" but of "steps taken or envisaged to implement the Convention".

The Kyoto Protocol, adopted in 1997, adds new commitments for Annex I Parties and confirms the general commitments from the Convention for non-Annex I Parties without modifying them in any significant respect.

With the Kyoto Protocol, Annex I Parties agreed to reduce aggregated emissions of carbon dioxide ($CO_2$), methane ($CH_4$) and nitrous oxide ($N_2O$), hydrofluorocarbons (HFCs), perfluorocarbons (PFCs) and sulphur hexafluoride ($SF_6$) in the period 2008 to 2012 relative to 1990 levels. Annex I Parties have individual limitation or reduction targets, as shown in Table 1.

To a certain extent, countries can reach their targets by trading emissions allowances with other countries or by implementing emissions reduction projects in other Annex I countries (via Joint Implementation or JI) or in developing countries, which do not have quantified targets themselves (via the Clean Development Mechanism or CDM).

Countries may also choose to implement the commitments jointly as a group (i.e. 'bubble'). The European Union has chosen to do so and has internally negotiated revised national targets that will be the basis for the assessment of their individual compliance with the Kyoto Protocol (see also Table 1).

Since the text of the Kyoto Protocol left some questions unanswered, the COP agreed on a substantive package clarifying the conditions of the implementation of the Kyoto Protocol in Marrakech in 2001, the so-called 'Marrakech Accords' (UNFCCC, 2001).

Entry into force of the Kyoto Protocol required ratification by 55 Parties including Annex I Parties responsible for 55% of the Annex I $CO_2$ emissions in 1990. Since the US (responsible for 36%) rejected the Kyoto Protocol in 2001, it was the Russian Federation (responsible for 17%) that had the decisive vote.[1] In October 2004, Russia ratified the Kyoto Protocol, which made the instrument enter into force on 16 February 2005.

## 3. History of the negotiations on a future agreement

The Convention states in Article 4.2(a) and (b) that Annex I Parties shall adopt and implement policies and measures to return their greenhouse gas emissions in 2000 to 1990 levels. A review of the adequacy of those paragraphs was called for to take place at the first session of the Conference of the Parties (COP 1) and a second review no later than 31 December 1998, and thereafter at regular intervals determined by the COP, until the objective of the Convention is met.

Figure 2 presents a timeline of past and future steps in the discussion on commitments, which are summarised below.

---

[1] For an analysis on the voting power of countries to bring the Protocol into force, see Wagner & Höhne (2001).

*Figure 2. Timeline of the negotiations on a future agreement*

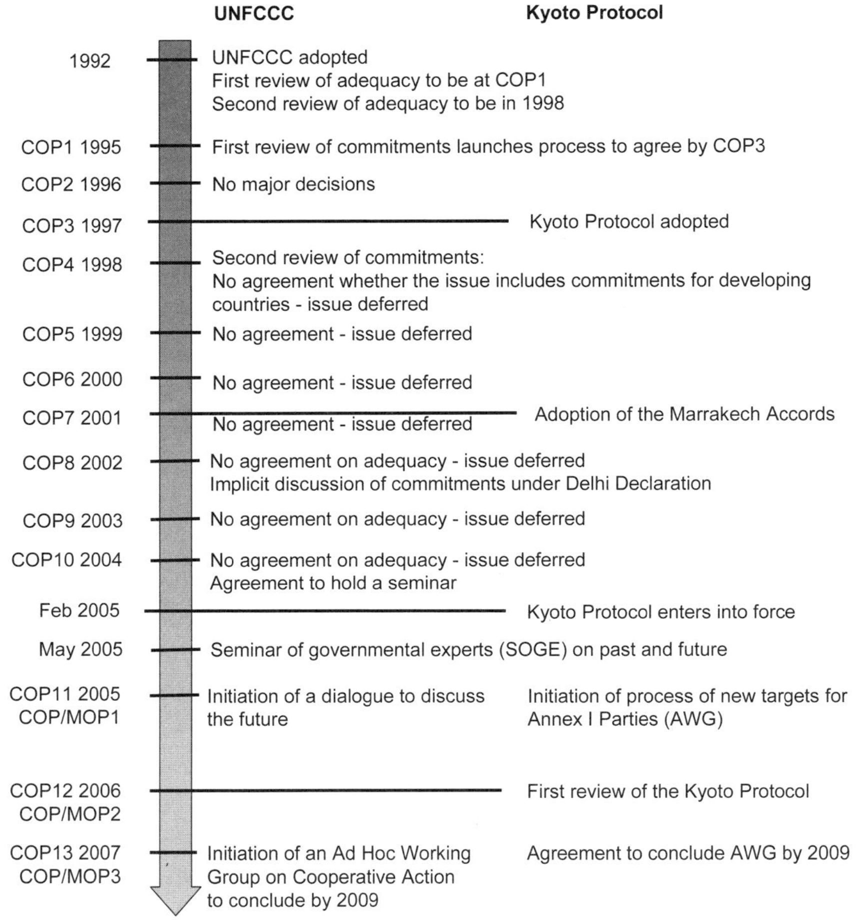


The first review at COP 1 (1995) concluded that the paragraphs in Article 4.2(a) and (b) were not adequate. With the 'Berlin Mandate', the COP initiated a process to strengthen the commitments of Annex I Parties without introducing any new commitments for non-Annex I Parties. The negotiations of the Ad Hoc Group on the Berlin Mandate (AGBM) resulted in the adoption of the Kyoto Protocol and its binding quantified targets for Annex I Parties at COP 3 in 1997. Although the subject of developing country commitments continued to be raised in the AGBM and at Kyoto itself, with a suggestion by some developed countries for a 'Kyoto Mandate' comparable to the Berlin Mandate, deep political divisions meant such proposals were not agreed.

The second review of adequacy was discussed at COP 4 in 1998, overshadowed by the issue of a future mandate that would allow discussion of additional developing country commitments. No agreement was reached. At COP 5 in 1999, again no agreement was reached and the agenda item was deferred in the following year to COP 6. Due to more urgent matters relating to Kyoto's unfinished business and the absence of any movement in Parties' positions, again no agreement was reached.

The text of the Kyoto Protocol, adopted at COP 3, left several details of its implementation open, e.g. how emissions from land use change and forestry would be counted. The Marrakech Accords, adopted at COP 7 in 2001, prescribe detailed rules for the implementation of the Kyoto Protocol. The completion of the Kyoto 'rulebook' provided an opportunity for raising the issue of future climate policy, but the uncertainty caused by the US withdrawal from the Protocol in March 2001 confounded political expectations. No one wanted to raise the deeply divisive issue given that the only future issue of concern to everyone was much more immediate: whether the Protocol would enter into force given that the US had pronounced it 'dead' and the double trigger required not only 55 countries but also coverage of 55% of Annex I emissions.

At COP 8 in 2002, the second review of the adequacy of Annex I Parties' commitments could not be resolved and was again deferred. However, the issue of a mandate that would allow discussion/negotiations of developing countries' commitments was raised in the context of the negotiations on a Delhi Declaration, which was to become the main outcome of the conference. The final Delhi Declaration, however, does not refer to the future. It was welcomed and supported by the G77 and the US, while the EU and Japan and Canada and the Central Group-Eleven (CG11, comprising most economies in transition included in Annex I) voiced their disappointment.

At COP 9 in Milan in 2003, all countries were silent on commitments after 2012 in the official negotiations. The EU did not want to repeat the negative experience of COP 8. But on the margins and at side events, the issue of commitments post-2012 was intensively discussed.

The COP 10 in Buenos Aires in the following year saw a fresh attempt to start official negotiations on commitments and actions beyond those set out in the UNFCCC/Kyoto. The ratification of the Kyoto Protocol by Russia just prior to the Conference, ensuring Kyoto's entry into force on 15 February 2005, gave a considerable political boost. Text was agreed to hold a 'seminar' on the future *and* the past, "bearing in mind that this seminar does not open any negotiations leading to new commitments" (UNFCCC, 2004).

The Kyoto Protocol's entry into force served as the background to the Seminar of Governmental Experts (SOGE), held in May 2005. Although it was a one-off event and informal in nature, the SOGE proved that sufficient political will existed amongst a sufficiently wide range of Parties to commence consideration of the legal nature and scope of additional future actions.

The Kyoto Protocol itself demands review of existing commitments within two articles. In Article 3, on the quantified commitments for Annex I Parties, the review of commitments inscribed in Annex B of the Protocol for Annex I Parties shall be initiated in 2005. Article 9 of the Kyoto Protocol calls for a general review of the Protocol coordinated with the review of the Convention, starting at the second meeting of the Parties to the Protocol, held in 2006.

COP 11 (2005) in Montreal was also the first Meeting of the Parties to the Protocol (COP/MOP 1). Two processes with relevance to a future agreement were agreed. The review of commitments for the post-2012 period for Annex I Parties inscribed in Annex B of the Protocol was initiated. An open-ended, Ad Hoc Working Group (AWG) was formed. These discussions will be undertaken only by the countries that ratified the Kyoto Protocol, i.e. excluding the US and Australia.[2] In addition, a two-year process was launched to discuss "long-term cooperative action to address climate change". Such action can commence any time and thus the process is not tied to the 2008-12 period. The process took the form of a dialogue held under the UNFCCC and therefore included the US and Australia. The mandate speaks of sustainable development, adaptation, technology potential and market-based opportunities as topics, but does not speak explicitly of emissions reductions. The mandate also states that it "will not open any negotiations leading to new commitments".

At COP 12, which is also COP/MOP 2, in November 2006, the first review of the Kyoto Protocol was initiated and completed without substantial conclusions. The discussion was deferred by the decision that the second review will take place in 2008. The review "shall not lead to new

---

[2] Australia ratified the Kyoto Protocol in November 2007.

commitments for any Party". It decided to consider the scope and content of the second review already in 2007.

In Bali in December 2007, COP 13/MOP 3 initiated the Ad Hoc Working Group on Long-term Cooperative Action under the Convention to finalise its work by COP 15, i.e. by the end of 2009. This group is to discuss "mitigation commitments or actions" by all developed countries and "mitigation actions" by developing countries. The conference also agreed that the ad hoc working group (AWG) under the Kyoto Protocol would complete its work in December 2009 too.

## 4. Current activities relevant to a future international agreement

The UNFCCC is the primary forum for discussion on climate change, but the scope and timing of a future agreement to combat climate change at the international level is also being discussed in other political fora. Figure 3 includes a few of these, which are discussed below. It is expected that in the long run, deliberations in all of these fora will lead to a more effective future international system to combat climate change.

*Figure 3. Overview of different strands of activities relevant to future international climate policy*

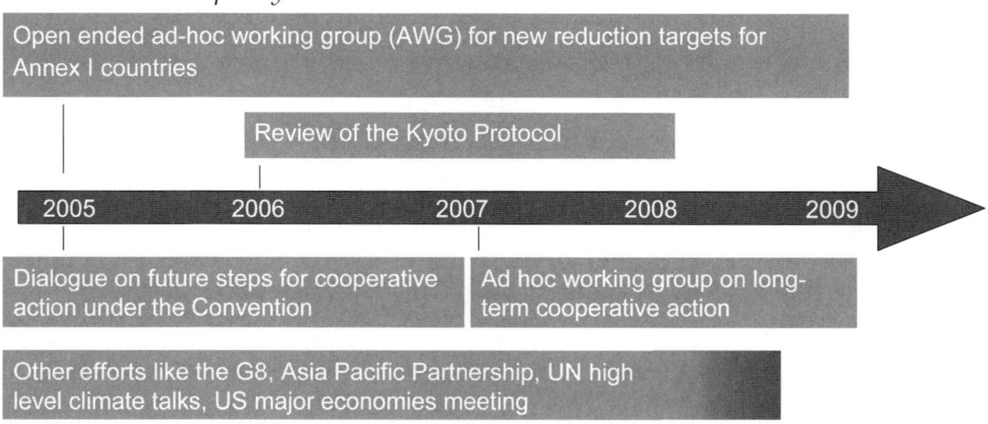


### 4.1 *Activities within the UNFCCC*

Many of the activities within the UNFCCC process are relevant for the development of the future agreement. The most important ones are discussed in more detail below and include:

- New commitments under the Kyoto Protocol and
- Further action under the Convention.

### *4.1.1 New commitments under the Kyoto Protocol*

The first Meeting of the Parties to the Protocol (COP/MOP 1, 2005) initiated the process of establishing commitments for the post-2012 period to be inscribed in Annex B of the Protocol for Annex I Parties. An open-ended, ad hoc working group was formed (AWG). A deadline for its completion was not fixed. These discussions are under the legal authority of Kyoto Parties with non-Parties such as the US having observer status.

At its first meeting in May 2006, the AWG did not decide a deadline for decisions, but stated that it should be in time so that there is no gap between the first and second commitment periods (UNFCCC, 2006). The group is now entering an analysis phase where countries are invited to present information relevant to reduction commitments, including mitigation potentials.

At COP/MOP 2 in November 2006, the AWG focused on the development of a work plan and a schedule for further meetings. The future work programme will include:

a) analysis of mitigation potential and ranges of emissions reduction objectives,

b) analysis of possible means to achieve mitigation objectives and

c) consideration of further commitments.

During its third session in May 2007, a roundtable took place to provide delegates an opportunity to discuss the current status of the scientific understanding and relevant experience.

At the first part of the fourth session in Vienna in August 2007, the AWG agreed for the first time an overall ambition level by recognising the findings of the Intergovernmental Panel on Climate Change (IPCC) that emissions of global greenhouse gases need to peak in the next 10 to 15 years and be reduced to very low levels, well below half the level in 2000 by the middle of the 21st century. It also noted that Annex I Parties as a group would need to reduce emissions in a range of 25-40% below 1990 levels by 2020. Taking these ranges as "useful initial parameters for the overall level of ambition", the AWG turned to the next item of its work plan, i.e. the means to achieve the mitigation objective.

In December 2007, the AWG agreed a detailed work programme to be able to conclude its work by the end of 2009.

COP/MOP 2 had agreed to have a second review of the Kyoto Protocol in 2008 and to discuss its scope in 2007. In 2007 the elements of the review were agreed. The decision still states that "the second review shall not lead to new commitments for any party".

### 4.1.2 *Further action under the Convention*

A two-year 'dialogue' process was launched at COP 11 (2005) to discuss "long-term cooperative action to address climate change". This process took the form of an open-ended dialogue by the COP itself, including the US and Australia. The mandate speaks of sustainable development, adaptation, technology potential and market-based opportunities as topics, but does not explicitly refer to emissions reductions. The mandate also states that the dialogue "will not open any negotiations leading to new commitments" but the two co-facilitators of the dialogue will report to COP 12 and COP 13.[3]

At the first meeting of the Convention dialogue, held in Bonn, 15-17 May 2006, Parties exchanged their views openly. Countries emphasised which topics should be the focus of the discussions. The theme of positive incentives for action in developing countries was heard many times.

During the second workshop of the Convention dialogue during COP 12 in November 2006 in Nairobi, several topics were presented and discussed. These included e.g. the Stern Review on the Economics of Climate Change, the World Bank Investment Framework on renewable energies, positive incentives to reduce emissions from deforestation in developing countries and integrating climate change into development strategies. Discussions were held e.g. on advancing development goals in a sustainable way and realising the full potential of market-based opportunities.

The third workshop of the Convention dialogue took place in Bonn in May 2007. It focused on realising the potential of technology and addressing action on adaptation. The fourth workshop in Vienna in August 2007 focused on financial flows necessary to slow climate change and on a possible future process. No formal decisions were adopted. The co-facilitators reported to COP 13 (2007).

---

[3] See Decision 1/CP.11.

At COP 13/MOP 3 in Bali, a process was agreed to negotiate a global climate change agreement by the end of 2009. The AWG on long-term cooperative action under the Convention shall complete its work for adoption by COP 15. The work programme will be developed at the first meeting and the aim is to "enable the full, effective and sustained implementation of the Convention through long-term cooperative action, now, up to and beyond 2012." A "shared vision including a long-term goal for emission reductions" will be addressed taking into account common but differentiated responsibilities and respective capabilities. Four major elements will be considered: mitigation, adaptation, technology and financing.

During the many discussions, the developing countries making up the G 77 and China made clear that the two tracks (the long-term cooperative action and the AWG) must remain separate and ensured that there was little discussion of their linkage. This was achieved and both the AWG and the long-term cooperative action under the Convention will continue in parallel under separate leadership. China also strongly opposed the Russian proposal on procedures for the approval of voluntary commitments (the third track). Russia was encouraged to continue discussion of its proposal under Article 9 and the AWG on long-term cooperative action.

One of the major decisions was the agreement that funding for adaptation projects in developing countries should begin under the supervision of the newly established Adaptation Fund Board. This should enable a rapid start for the Adaptation Fund (which is funded by a 2% levy on CDM projects and has currently reached €37 million with an increase expected to an estimated $80-300 million in the period 2008-12 with projects in the CDM pipeline). There was no agreement on additional practical adaptation measures and the issue will be re-visited at SBSTA[4] in June 2008.

Another of the big successes at Bali was the adoption of a pilot work programme for further methodological work to reduce emissions from

---

[4] Subsidiary Body for Scientific and Technological Advice, established at the first Conference of the Parties (COP) to the UNFCCC in August 1995 to provide the COP and, as appropriate, its other subsidiary bodies with timely information and advice on scientific and technological matters relating to the Convention.

deforestation and forest degradation. This programme will focus on assessments of changes in forest cover and associated greenhouse gas emissions, methods to demonstrate reductions of emissions from deforestation and the estimation of the amount of emissions reductions from deforestation. Demonstration activities and capacity-building will be undertaken. Norway offered considerable funding to support this decision.

## 4.2 *Activities outside the UNFCCC*

Outside the UNFCCC, several processes with relevance to a future climate agreement have been initiated. Four are of particular relevance: the Gleneagles G8 discussions on climate change, the Asia-Pacific Partnership (AP6), the United Nations high-level climate change talks and the US major emitters initiative.

### *4.2.1 Gleneagles G8 plus 5 process*

During the G8 meeting 2005 in Gleneagles, Scotland, at which five developing countries - Brazil, China, India, Mexico and South Africa - participated, the Gleneagles Communiqué and Plan of Action on Climate Change, Clean Energy and Sustainable Development was released. The G8+5 group emphasise the need to stop and reverse the increase of greenhouse gas emissions. It includes three areas of future work: the ministerial dialogue, cooperation with the International Energy Agency (IEA) and with the World Bank.

*Ministerial dialogue*

A major commitment of the G8 Summit in Gleneagles was to "take forward a Dialogue on Climate Change, Clean Energy and Sustainable Development, and to invite other interested countries with significant energy needs to join". This Gleneagles Dialogue is an informal forum for discussion. Its objective is to complement and reinforce the formal negotiations within the UNFCCC by trying to create the conditions necessary for successful agreement.

The dialogue encompasses 20 countries (G8+5 together with Australia, Indonesia, Iran, Nigeria, Poland, South Korea and Spain), the European Commission and key international organisations including the World Bank and the IEA. The UNFCCC Secretariat also participates.

The July G8 Summit in Russia (2006) included a series of meetings with much of the focus on energy security and access to supplies.

Nevertheless, the Russian Summit reaffirmed the G8's commitments to meet the objectives of reducing greenhouse gas emissions and of dealing with climate change, including through promoting an inclusive dialogue on further action in the future. The second ministerial dialogue meeting was hosted by Mexico in October 2006 (UK Defra, 2005).

Under the German Presidency, the G8 continued working on the Gleneagles Plan in 2007. The G8 countries committed to reduce global $CO_2$ emissions by 2050 by at least 50%. This aim shall be reached within the UN process and in cooperation with emerging economies. It agreed to finalise an agreement of the large emitters by 2008 and a global agreement under the UNFCCC by 2009.

The 2008 G8 Summit, presided over by Japan, will conclude the G8 process on climate change with a final report on previous work under the dialogue being submitted for the consideration of G8+5 leaders in Japan.

*IEA*

For the future cooperation between G8 and IEA, the Gleneagles Communiqué describes the IEA as advisor "on alternative energy scenarios and strategies aimed at a clean, clever and competitive energy future" (UK Presidency of the G8, 2005).

As part of this work, the IEA has published a major new report called *Energy Technology Perspectives: Scenarios and Strategies to 2050* (IEA/OECD, 2006). Another publication in this context focuses on energy efficiency in industry (IEA, 2007). The IEA's key findings will be delivered at the G8 Summit in Japan in 2008 (IEA, 2005).

*World Bank*

In the Gleneagles Communiqué, the participants describe the future role of the World Bank as taking "a leadership role in creating a new framework for clean energy and development, including investment and financing" (UK Presidency of the G8, 2005).

International financial institutions can be an important source of finance, policy and technical advice regarding the financing of investments needed for clean energy development. Also the creation of new financing instruments could support this effort in several possible ways (World Bank, 2006):

- *Clean Energy Financing Vehicle (CEFV).* This could provide a mechanism to transfer high-efficiency technology by blending grants and carbon finance.
- *Power rehabilitation financing facility.* Failures of supply can cause high costs. This facility could enable developing countries to rehabilitate inefficient plants without loss of power.
- *Project Development Fund.* 'Bankable' projects seem to be needed. Such projects with participation of the public and the private sector could be addressed with this fund.
- *Venture capital funds for technology adoption.* These funds could finance the development, adoption and penetration in the market of promising, new and clean technologies.

The G8 process has created new momentum for the international discussions on climate change. It has raised the issue to the level of heads of state and given new direction to existing institutions such as the IEA and the World Bank. It is seen as a process that can reintegrate the US and at the same time engender a constructive dialogue with the largest developing countries. By creating the conditions under which any future agreement could be successfully implemented, the G8 activities can complement the UNFCCC process.

### *4.2.2 Asia-Pacific Partnership (APP)*

The Asia-Pacific Partnership on Clean Development and Climate (APP), also known as AP6, was initiated by Australia, China, India, Japan, South Korea and the US (see http://www.asiapacificpartnership.org), and recently joined by Canada. The participating countries first met under this non-legally binding framework in January 2006.

The purposes of this partnership are to "advance clean development and climate objectives [...] The Partners will enhance cooperation to meet both [... their] increased energy needs and associated challenges, including those related to air pollution, energy security, and greenhouse gas intensities, in accordance with national circumstances."

The six founding countries represent about half of the world's economy, population and energy use. Globally, they are important producers of coal (65%), cement (61%), steel (48%) and aluminium (35%) (AP6, 2006).

The main institution of the AP6 is the Policy and Implementation Committee, which is chaired by the US. It is responsible for management of the implementation of the cooperative activities of the partnership and its task forces. The Administrative Support Group, which coordinates the communication, is also established by the US. Eight government-industry task forces have been set up, focusing on power generation and key industry sectors of the partner countries: cleaner fossil energy (co-chaired by Australia and China); renewable energy and distributed generation (co-chaired by Korea and Australia), power generation and transmission (co-chaired by the US and China); steel (co-chaired by Japan and India); aluminium (co-chaired by Australia and the US); cement (chaired by Japan); coal mining (co-chaired by the US and India) and buildings and appliances (co-chaired by Korea and the US). The transport sector is not covered. The task forces will formulate action plans that outline short-and medium-term action necessary to reach the Partnership's aims.

The task forces shall "drive improvements with regard to best practices and ensure that a range of technologies is developed and repeatedly demonstrated so that scale is increased and costs are reduced" (AP6, 2006).

The funding of this partnership is voluntary. The US pledged to contribute up to $50 million in 2007 (US OMB, 2006). The Australian Government plans to invest a further $100 million in Australian dollars (roughly $75 million) over five years (Australian Government, 2006).

The economic research agency of the Australian government concluded that the efforts from the Asian-Pacific Partnership could reduce global emissions from 22 GtCeq (gigatonnes of carbon equivalent) in 2050 under a reference case to 17 GtCeq, i.e. emission reductions of 23% compared to the reference case (ABARE, 2006). The WWF criticises this as being far too little because the emissions would still lead to a global temperature increase of 4°C (WWF, 2006). A target of 2°C would require global emissions to be below 1990 levels (7 GtCeq) in 2050.

The Asia-Pacific Partnership is seen by some of its members as an alternative to the UNFCCC and Kyoto process and by others as a complement to it. Its impact alone does not seem sufficient to keep the average global temperature increase below 2°C, which is the goal for some countries. It also still has to prove to be operational as a new institution. But so far, its existence has not significantly influenced the UNFCCC process. Indeed, it could complement it well, if it focuses on the development of particular technologies.

### 4.3 *United Nations High-Level Climate Change Talks*

UN Secretary General Ban Ki-moon announced in May 2007 that he hoped to bring the world's leaders together to discuss a future agreement on curbing greenhouse gas emissions "as recognition of the fact that climate change needed to be addressed at a higher level than that of environmental ministers, because the issue also had developmental and economic impacts". The meeting, on 24 September 2007, coincided with the UN General Assembly. At least 100 countries attended, over half of which represented by heads of state. The meeting provided impetus for COP 13 to adopt a mandate for a future climate agreement in Bali, Indonesia in December 2007.

### 4.4 *The US major economies meetings*

In May 2007, just before the G8 summit in Germany, US President George W. Bush announced a new initiative, to bring together the largest emitters to agree on a framework for future action on climate change by the end of 2008. Under the initiative, each of the major emitting countries would establish its own greenhouse gas emissions targets, goals and programmes according to national circumstances. The pledges would be reviewed regularly.[5] Prior to the first meeting on 27 September 2007, the initiative was re-named the major economies meeting on energy security and climate change.

France, Germany, Italy, the UK, Japan, China, Canada, India, Brazil, South Korea, Mexico, Russia, Australia, Indonesia, South Africa and the UN participated in the first meeting, held in Washington, D.C., on 27-28 September. The invitation letter stated that: "The United States is committed to collaborating with other major economies to agree on a detailed contribution for a new global framework by the end of 2008, which would contribute to a global agreement under the UN Framework Convention on Climate Change by 2009."

---

[5] See US White House (2007).

The timing of this initiative raised speculation that it was intended to distract attention from the G8 and the UNFCCC process. Many observers commented that such a voluntary pledge and review process could hardly achieve the significant reductions that would be necessary to limit global temperature increases to 2°C. This concern became secondary after the US agreed to the 'Bali Roadmap', and several additional meetings of the major economies initiative are planned during 2008.

## 5. Conclusions

Starting in 2005, a new momentum entered the negotiation process. After the entry into force of the Kyoto Protocol, climate change received high-level political attention in several parallel initiatives (UNFCCC, G8, APP, UN high-level talks, major economies meeting). With the start of two dialogues under the UNFCCC and the Kyoto Protocol in 2005, the G8 follow-up processes, the APP, the United Nations high-level meeting and the US major economies meeting, enormous momentum was built up for the COP in Bali 2007. Some argue that the large number of processes outside the UNFCCC could threaten the authority of the UNFCCC process as the international forum on climate change. But it seems that these processes have positively influenced the UNFCCC negotiations and have provided new input to them. And in the end, all processes influence each other and an agreement in one will most certainly play an important role in the other processes. The negotiating efforts will be intensified in the next two years, to reach a global agreement by the end of 2009.

Looking at the positions of countries, there are many gaps and areas of disagreement that must be bridged before any new universal agreement can be adopted. Countries will have positions that go further than they officially announce, but it is unlikely that all have solid positions on all topics necessary to form a full regime. The 'Bali Roadmap' intends to cover mitigation, adaptation, technology and financing.

The issue of ambition level has received some attention, but only a few countries have spoken openly about it. The EU and other countries are in favour of a 2°C target, The Alliance of Small Island States (AOSIS) calls it unacceptably high, but many countries have been silent on the issue of a long-term goal. Translating any long-term goal into global mid-term emission levels was only done by the EU, calling for at least 50% reduction of global emissions in 2050 below 1990 levels. Numbers for necessary reductions by 2020 by Annex I countries were only provided by the EU

(30%) and the environmental NGOs (30-35%). It is remarkable that the AWG has made a quite firm decision to aim for 450ppmv $CO_2$eq. and for reductions by Annex I countries 25-40% below 1990 levels in 2020. The mandate for the ad hoc group under the convention could not agree on such a goal.

All countries seem to agree that some developing countries should be given 'positive incentives' to participate, but no country has specified exactly how such positive incentives should be designed. No country has mentioned a list of countries. The EU and AOSIS have suggested that it should be determined on the basis of capability and responsibility.

All countries agree that adaptation will be a major element in a future climate treaty, but no Party has made concrete proposals on how it should be done. Here also the scientific policy-related literature has far fewer insights and proposals compared to mitigation.

In conclusion, it seems that the large number of discussion processes can lead to a more enhanced consideration of the issue of climate change at the international level. But the urgency of action that is suggested by climate science has yet to lead to accelerated analytical work that could feed into negotiations on the time scale needed to ensure that there is no gap between the first and second commitment periods. Progress now seems possible under the various future action-related processes and the 'Bali Roadmap' for reaching agreement by 2009. It remains to be seen whether such an agreement can be reached with sufficient stringency to avoid dangerous interference with the climate system.

## References

ABARE (Australian Bureau of Agricultural and Resource Economics) (2006), *Technological development and economic growth,* Research report 06.1 to the Inaugural Ministerial Meeting of the Asia-Pacific Partnership on Clean Development and Climate, Sydney, 11-13 January, 2006, Canberra, Australia (http://www.abare.gov.au/publications_html/climate/climate_06/06_climate.pdf).

AP6 Asia-Pacific Partnership (2006), Homepage of the Asia-Pacific Partnership (http://www.asiapacificpartnership.org/).

Australian Government (2006), "Asia-Pacific Partnership sets new path to address climate change", Press release of the Australian Government, Canberra.

Depledge, J. (2000), *Tracing the origins of the Kyoto Protocol: An article by article textual history, FCCC/TP/2000/2,* UNFCCC, Bonn (http://www.unfccc.int).

Höhne, Niklas (2006), *What is next after the Kyoto Protocol? Assessment of options for international climate policy post-2012,* Amsterdam: Techne Press (http://www.technepress.nl/publications.php?id=13).

IEA (International Energy Agency) (2005), *Climate change, clean energy and sustainable development – IEA's G8 Gleneagles programme* (http://www.iea.org/G8/index.htm).

IEA (International Energy Agency)/OECD (2006), *Energy Technology Perspectives 2006, Scenarios & Strategies to 2050,* Paris.

IEA (International Energy Agency) (2007), *Tracking Industrial Energy Efficiency and $CO_2$ Emissions,* Support to the G8 Plan of Action, Paris.

Meinshausen, Malte (2005), "On the risk of overshooting 2°C", paper presented at the International Symposium on Stabilisation of Greenhouse Gas Concentrations – Avoiding Dangerous Climate Change, 1-3 February, Exeter, UK (http://www.stabilisation2005.com/programme.html).

Oberthür, Sebastian and Hermann E. Ott. (1999), *The Kyoto Protocol – International Climate Policy for the 21st Century,* Berlin: Springer.

UK Department for Environment Food and Rural Affairs (Defra) (2005), *Gleneagles dialogue – meeting the challenge,* London (http://www.defra.gov.uk/news/2005/051102a.htm).

UK Presidency of the G8 (2005), *G8 Factsheet: Dialogue on climate change, clean energy and sustainable development* (http://www.g8.gov.uk).

UNFCCC (2001), Report of the Conference of the Parties on its seventh session, held at Marrakesh, 29 October-10 November 2001 (http://unfccc.int/resource/docs/cop7/13.pdf).

UNFCCC (2004), Report of the Conference of the Parties on its tenth session, held at Buenos Aires, 6-18 December, Part One: Proceedings, FCCC/CP/2004/10 (http://www.unfccc.int).

UNFCCC (2006), Planning of future work, revised report proposed by the Chair, FCCC/KP/AWG/2006/L.2/Rev.1 (http://www.unfccc.int).

US Office of Management and Budget (OMB) (2006), Budget of the Department of State and Other International Programs, Washington, D.C. (http://www.whitehouse.gov/omb/budget/fy2007/state.html).

(US) White House (2007), "President Bush Participates in Major Economies Meeting on Energy Security and Climate Change" (http://www.whitehouse.gov/infocus/environment).

Wagner, Fabian and Niklas Höhne (2001), "Influence of national governments for or against the entry into force of the Kyoto Protocol: A Banzhaf index analysis", *Climate Policy, 1.*

World Bank (2006), *Clean energy and development: Towards an investment framework,* Washington, D.C. (http://siteresources.worldbank.org/DEVCOMMINT/Documentation/20890696/DC2006-0002(E)-CleanEnergy.pdf).

WWF (2006), *Asia-Pacific Partnership sets world up for massive global warming,* World Wide Fund For Nature (http://www.panda.org/about_wwf/what_we_do/climate_change/news/index.cfm?uNewsID=56020).

Yamin, F. and J. Depledge (2004), *The International Climate Change Regime, A Guide to Rules, Institutions and Procedures,* Sussex and Cambridge: Institute of Development Studies/Cambridge University Press.

# 3. Getting Serious about Mitigation Potential: The Work of the Ad Hoc Working Group on Further Commitments for Annex I Parties under the Kyoto Protocol

*Adrian Macey**

## 1. Introduction

Prior to Bali, the only negotiations leading to binding commitments to reduce greenhouse gas emissions after 2012 are taking place in the Ad Hoc Working Group on Further Commitments for Annex I Parties under Article 3.9 of the Kyoto Protocol (AWG). The negotiations are being carried out in light of the overall objective of the UN Framework Convention on Climate Change (UNFCCC) of stabilising greenhouse gases in the atmosphere at a safe level, with developed countries required to continue to take the lead. The first stage of the AWG work programme involves analysis of mitigation potential,[1] which is seen as a key input to determining overall reduction ranges for Annex I parties and ultimately to individual national commitments. Before discussing the AWG's work on mitigation potential, it will be helpful to see where it is situated in the global context.

* Climate Change Ambassador, Ministry for Foreign Affairs and Trade, New Zealand.

[1] See document FCCC/KP/AWG/2006/4, http://unfccc.org.

## 2. The global context

The effort needed to reduce greenhouse gas emissions to safe levels is daunting, when business as usual projections of emissions for the next few decades are shown. Can the world achieve this ambitious task? Estimating the global mitigation potential is the beginning of the answer. If the potential is there, attention can turn to how it is realised. Translating the potential into action will require organisation, resources and international commitments, and will pose challenges to all economies in adopting low emissions pathways without compromising development.

The scale of the challenge is best understood if the figures are allowed to speak for themselves. In its Fourth Assessment Report, the Intergovernmental Panel on Climate Change (IPCC) estimates that the world will have to find about 30 gigatonnes of $CO_2$ equivalent reductions from baseline by 2030 to reach the lower stabilisation levels of 440-490 parts per million of $CO_2$ equivalent.[2] This means globally at least 50% below 2000 levels by 2050 or for Annex I Parties 25-40% below 1990 levels by 2020. This range is identified by the IPCC as most likely to limit the global temperature rise to between 2-3°C, beyond which there is an increased risk of severe impacts.

The Fourth Assessment Report contains an estimate of mitigation potential, which shows that such reductions are theoretically achievable, at a cost, without assuming any major technological breakthroughs. The IPCC makes use of both 'bottom up' and 'top down' studies to derive its figures, and concludes that there is increasing convergence of these two approaches. At two different prices of carbon ($50 and $100), both much higher than current prices, it estimates that Annex I countries could find between 6 and 11 gigatonnes reductions, and non-Annex I countries, between 6-19 gigatonnes.[3] At $50 the chances of achieving the target range would be marginal, but at $100 it could be exceeded. A note of caution

---

[2] These figures are from the presentation by Bert Metz, Co-chair of IPCC AR4 Working Group III to the AWG roundtable in Bonn, 12 May 2007 (htttp://www.ipcc.ch/unfccc_pdf/session3_metz.pdf).

[3] At its third session in May 2007, the AWG held a roundtable discussion covering these issues with several invited speakers (see footnote 2). Presentations are available on the UNFCCC website (http://unfccc.int/kyoto_protocol/items/3951.php).

should nonetheless be sounded as the IPCC points out that this "economic potential" is greater than the "market potential"[4] realisable under current conditions without changes to government policy settings.

Notwithstanding their limitations, the IPCC figures on the potential and costs of mitigation on the global level tend to support the broad findings of the Stern Review on the Economics of Climate Change,[5] that mitigation is achievable at cost levels around 0.1% lower per annum growth in world GDP, and at any rate far below the assumed long-term costs of inaction. It is clear, however, that such global targets can only be achieved by measures in a wide range of countries, and notably in the major economies that are responsible for the bulk of global emissions.

The figures in the Stern Review, and even more so those in the IPCC report because they are endorsed by governments, are helpful politically because they indicate that it is worth making the effort. The target is not out of reach if early action is taken. But this information is of little practical use by itself. The figures are not broken down by country and so have no operational value in the short to medium term, which is what is needed by Parties in the AWG negotiations. It should also be noted that the Annex I mitigation potential figures in the IPCC's report overstate what could be achieved post-2012 by present Kyoto Protocol Parties, by up to 30%, as they include both Australia and the US, which do not currently have commitments under the Kyoto Protocol.

There are also many uncertainties in the overall figures. The stimulus that could be given by an expanded and efficient global carbon market could be considerable. Technological advances beyond steady improvements that can be assumed in such areas as energy intensity are also impossible to predict. It is to be expected that the longer the period being addressed, the more the likelihood of major technological breakthroughs that could increase mitigation potential. Finally, no current calculations of mitigation potential take account of another set of factors – lifestyle changes – that is likely to have significance for countries pursuing sustainability policies for their economies and societies, or where the

---

[4] For an explanation of the different types of potential, see IPCC Third Assessment Report (2001), Working Group III, Chapter 5, especially Figure 5.1.

[5] Nicholas Stern, *The Economics of Climate Change*, Cambridge: Cambridge University Press, 2007.

population itself embraces a more sustainable lifestyle. For sectors such as transport, these lifestyle changes could be an important tool in emissions reduction.

There is a requirement for there to be no gap between the end of the first commitment period (2008-12) and the start of the second one, which gives the work of the AWG some urgency. The section of the AWG's work programme dealing with mitigation potential of Annex I Parties provides for the consideration of the "policies, measures and technologies" at their disposal, as a basis for determining possible ranges of emissions reductions. This is quite a broad sweep but necessary to the ultimate determination of specific commitments by country. Most notably, it enables national circumstances to be taken into account, reflecting the fact that there are large differences among the economies of Annex I Parties.

An understanding of mitigation potential is a necessary but not sufficient condition for determining commitments. Parties will want to see their future commitments in light of the global context of the overall post-2012 climate arrangements. Relevant factors will include whatever global reductions goal may be decided, the evolution of the science, the actions of Parties that do not presently have commitments, and indeed the ambition of new commitments taken on by other Parties that already have commitments. The present Kyoto Protocol negotiations cannot deliver a comprehensive post-2012 agreement, and it is implicit in the AWG's work that Annex I Parties alone cannot deliver the stabilisation objective of the Convention. There is no provision for involving other countries in a second commitment period, despite references to a "shared vision" and to the overall objective of the Convention. Some Parties may judge that they need to make future commitments conditional on wider uptake of mitigation measures. The EU for example has made its level of ambition contingent on participation and comparable efforts by others, and its expectations extend beyond Annex I countries.

## 3. How should mitigation potential be determined?

The commonsense understanding of mitigation potential as what each country is capable of doing to reduce GHG emissions does not address the issue of the basis for comparison. There is no agreed definition of mitigation potential, still less of the method for determining it. Mitigation potential can be viewed from several perspectives, including market potential, economic potential, socio-economic potential, technological potential and physical potential.[6] Should account be taken of simply economic and technological factors, i.e. the cost and availability of technologies? Should it include the possibility of lifestyle changes? To what extent should it include the possibility of countries changing the focus of their economies, for example to exit relatively emissions-intensive activities? If this simply leads to carbon leakage, a geographical transfer of the same emissions, it is not net mitigation from the perspective of the planet. Mitigation also has co-benefits that can reduce net costs - should the domestic costs be therefore reduced and mitigation potential correspondingly increased? Ideally, the answer would be yes, but this could require cost-benefit analysis for which there do not appear to be adequate tools at present. Finally, it would probably be advisable from the same point of view of clarity and comparability to treat mitigation potential separately from the use of sinks.

A first distinction that will be useful in the Kyoto Protocol negotiations is that between domestic and total mitigation potential. There are in fact two types of distinctions possible here. First, between domestic potential realisable entirely within national borders and the greater potential at similar cost using international mechanisms such as emissions trading or the Clean Development Mechanism. Should estimates of mitigation potential be increased to take account of the possibility of using the international market to reduce costs? At the first stage this is probably not useful, especially as there is no certainty about the state of the international carbon market after 2012. There may also be a risk of double counting - for example a Joint Implementation project reducing emissions in one Annex I country paid for by a business in another. The second possible distinction is between domestic mitigation potential and a 'responsibility target', which would include an additional quantity of emissions reductions, unrelated to domestic mitigation potential that would be undertaken as part of a country's contribution to the global effort. This could take place in third countries through Clean Development Mechanism projects or indirectly through purchasing emissions units on the market. One might expect there to be a correlation between this additional element and ability to pay, as measured for instance by GDP.

---

[6] IPCC (2001), op. cit.

This discussion will concentrate on domestic potential, which is important to get right as a baseline for the other dimensions, and which also provides useful comparative information among countries. Rather than seek a definition, it will be more instructive to look at the factors that are commonly used in discussions of mitigation potential. Domestic mitigation potential should initially be an objective statement about what is possible to achieve, and at what cost, not what should be done. The 'is' must be established before getting to the 'ought'. Mitigation potential should thus be distinguished from such concepts as *responsibility* for past emissions and *capability*, or ability to pay, important as these factors may be in the determination of final targets or commitments.[7]

Whatever the base year taken, the actual costs that Parties will incur are through mitigation action applied to future emissions. The base year 1990 used in the Kyoto Protocol is a logical point of reference for Parties with existing commitments. But given the strong divergence of growth of economies and emissions since 1990, a base year closer to the present day would be more useful as a point of reference in order to compare mitigation potential among countries. Having some comparative mitigation potential information will be useful in building confidence that one's country is being fairly treated and conversely, that others are making a contribution commensurate with their ability to do so. This fair burden-sharing aspect is likely to be increasingly important in the final stages of the negotiations.

## 4. AWG discussions

The UNFCCC secretariat completed a 'technical paper' in July 2007 (UNFCCC, 2007), for the AWG to help in the assessment of mitigation potential. This is a useful survey of relevant factors and indicators (the term 'criteria' in its mandate from the AWG was judged inappropriate because of the risk of compromising the secretariat's neutrality). The factors and indicators are brought together from the literature on the subject as well as from submissions from Parties and presentations to the UNFCCC. The

---

[7] For a discussion of the three concepts of potential, responsibility and capability, see Matthew Ogonowski, "Comparability and Economics of Developed Country post-2012 Mitigation Commitments", July 2007 (http://www.ccap.org/international/July%202007.htm).

Secretariat does not advocate any one approach or set of factors, but few of these indicators are controversial and there appears to be a degree of common understanding. So despite its technical nature, the paper will remain very useful to inform policy beyond its short-term purpose in the AWG negotiations, and would be worth developing further.

The paper discusses both nationwide and sectoral indicators. The nationwide figures include per capita emissions, which are often suggested as the most reliable and equitable indicator for the future. However this shows an average of 11 tonnes $CO_2$ equivalent per capita among Annex I countries, with a variation of 3.0 to 27.7 tonnes, so is unlikely to be useful on its own without reference to the underlying sectors. Per capita emissions could certainly be useful in the very long term, but as a guide to action in the short to medium term they are of little use. They can also be misleading. A high per capita figure may not necessarily indicate an inefficient economy with high mitigation potential. For example, to take the case of New Zealand, where 49% of GHG emissions come from methane and nitrous oxide from agriculture, there is no available mitigation technology that can currently make an impact, especially on methane from enteric fermentation. New Zealand's per capita emissions therefore appear relatively high despite its agricultural sector being close to or at world's best practice in terms of greenhouse gas intensity. Similarly, a heavily trade-dependent economy whose exports are more emissions-intensive than its imports may also have above-average per capita emissions.

A further objection to using per capita emissions as the predominant metric is that it carries a connotation of a right to emissions per se rather than a right to growth, whereas the focus of the international community should be on a right to growth, but along a low-emissions pathway.

Other nationwide indicators also show wide variation, for example, estimated population growth (a key push factor for emissions) between 2004 and 2020 varies from –16% to +20%. GHG emissions trends themselves show a huge variation between 1990 and 2004, from –60% to +73%. Such variations are a good illustration of why no single metric would give an accurate representation of mitigation potential.

National circumstances are best described on a sectoral basis, because aggregates can mask some key differences among economies. The UNFCCC paper shows variations in the sectoral indicators to be at least as great as for national indicators. A key indicator of mitigation potential is the share of renewable energy in electricity production. This has an average value of 21.3%, but shows huge variations between negligible and 100%. Given the large proportion of global emissions growth coming from the energy supply sector, this is a highly important indicator. Mitigation potential in the energy sector will be much lower for countries with a high proportion of renewables and will greatly affect national aggregate emissions reductions commitments that they can take on. These countries will have to look for gains in energy efficiency.

Given that no single indicator of potential is going to prove sufficient, it would be theoretically possible to agree on a combination of indicators with appropriate weightings for each and expressed as a mathematical formula, to be applicable to all economies. It is unrealistic to expect this to be negotiable, as each country would naturally seek to gain advantage from the weightings. A sectoral breakdown of mitigation potentials (which could be aggregated to give a national figure) will give the most informative basis for cross-country comparisons. For the present, reference to both nationwide and sectoral indicators is appropriate. One broad metric that would be useful would be sectoral abatement cost curves showing costs per tonne of $CO_2$ equivalent GHG reduction, but this information is not widely available yet. Nor are accurate figures available for the macroeconomic effects of mitigation action at the country level.

One conclusion that can be drawn from the albeit incomplete figures available so far is that after 2012, there is likely to be a much larger spread around an average figure for GHG reduction commitments than in the first commitment period. This information was not available at the time of the negotiation of the first commitment period. For European Union members, national and sectoral differences can be accommodated within a single overall EU target for the purposes of the negotiation. While this flexibility is theoretically available to other Annex I members under Article 4.1 of the Kyoto Protocol, it does not seem a practical proposition. So whereas the EU is able to agree to a target and negotiate internal burden-sharing later, that negotiation will have to take place earlier for non-EU Annex I countries, whose economies are as diverse as those of EU members.

Mitigation potential is dynamic because it will evolve through time. This fact should be taken into account in the post-2012 arrangements. If potential were to increase, countries could take on more ambitious targets. Evolution in the type of economic activity - for example moving from manufacturing to a service economy, or technological breakthroughs - could increase the mitigation potential. To return to the New Zealand example of agricultural emissions, with current technologies, there has been a gradual improvement in methane intensity from livestock, but if the combined stimuli of the proposed emissions trading scheme[8] and increased international research efforts[9] were to lead to a breakthrough on enteric fermentation, New Zealand might be able to increase its domestic mitigation efforts. Strong economic growth could increase affordability of mitigation but will also tend to increase net emissions.

This argues for post-2012 arrangements having built-in flexibility, unlike the first commitment period where no change was possible even though the commitments were negotiated over ten years before they became effective. The flexibility would need to be limited, because some stability is necessary to allow for the efficient functioning of carbon markets. Flexibility could be achieved in various ways - regular review, breaking down targets by sectors, sectoral agreements, some form of adjustment mechanism according to changes in economic or technological factors. Before a global agreement is in place, post-2012 targets may well be provisional - this would give the market a positive signal, without reducing flexibility pending final decisions.

In a context where uncertainties remain, the concept of an 'iterative' process used by the AWG is useful. There is unlikely to be a simple linear progression through the work programme from potentials through reduction objectives and available means to establishing final commitments. There is incomplete information at all stages of the process, including some of the underlying science, so it may be necessary to revisit elements that have already been traversed once or more by the AWG, before reaching final decisions.

Beyond the AWG, estimates of mitigation potential using these national and sectoral indicators could be equally useful in the case of voluntary GHG reduction action by non-Kyoto Parties or developing countries, or different forms of commitment other than the national

---

[8] In September 2007, the New Zealand Government announced a proposed emissions trading scheme which would progressively be applied to all sectors and all gases. Agriculture would be included from 2013.

[9] New Zealand is leading the establishment of an international research network on livestock greenhouse gas emissions.

'quelros'[10] of the first commitment period, through sectoral agreements for example.

## 5. Conclusion

Addressing mitigation potential is a necessary step in the work of the Kyoto Protocol Parties towards establishing commitments after 2012 and could also have a wider application in a comprehensive post-2012 agreement. Final decisions about global goals, commitments and burden-sharing will be decided through negotiations where equity and efficiency will be important considerations. Informed discussions about mitigation potential, including economic modelling and comparisons across different economies, will enable those decisions to be better-informed and more transparent. Government and stakeholders already have more information at their disposal than at a comparable time in the negotiations of the first commitment period. This work will also be valuable in the longer term as potentially the Kyoto Protocol negotiations become more closely linked with negotiations under the Convention.

## References

Ad Hoc Working Group on Further Commitments for Annex I Parties under the Kyoto Protocol (2007), presentations at roundtable discussion, May (http://unfccc.int/kyoto_protocol/items/3951.php

Intergovernmental Panel on Climate Change (IPCC) (2001), Third Assessment Report, Working Group III.

Metz, Bert (2007), presentation by Co-chair of IPCC AR4 Working Group III to the AWG roundtable, Bonn, 12 May (htttp://www.ipcc.ch/unfccc_pdf/session3_metz.pdf).

Ogonowski, Matthew (2007), "Comparability and Economics of Developed Country post-2012 Mitigation Commitments", July (http://www.ccap.org/international/July%202007.htm).

Stern, Nicholas (2007), *The Economics of Climate Change*, Cambridge: Cambridge University Press.

UNFCCC Secretariat (2007), *Synthesis of information relevant to the determination of the mitigation potential and to the identification of possible ranges of emission reduction objectives of Annex I Parties*, Technical Paper, FCCC/TP/2007/1, July.

[10] Quantified emission limitation and reduction objectives.

# 4. More Inconvenient Truths: Towards an Equitable Global Climate Change Regime

*Surya P. Sethi*[*]

The international debate on climate change is not placing sufficient emphasis on the issue of equity while seeking a post-Kyoto regime. Sir, Gordon Brown, the British Prime Minister, underscored the need for an equitable global climate regime when he stated: "Climate change is an issue of justice as much of economic development. It is a problem caused by the industrialised countries whose effects will disproportionately fall on developing countries".[1] This statement should be the beacon for those framing the post-Kyoto regime.

Unfortunately, the internationally quoted body of work on climate change typically combines scientific uncertainty with controversial value choices to estimate the social cost of carbon and the expected cost of catastrophic events resulting from high levels of greenhouse gas concentrations. It then goes on to do a cost-benefit analysis to choose the most economically viable, and, one might add, politically palatable though less-constraining, mitigation trajectory. Finally, it deftly finesses the implicit high probability of setting in motion critical non-linear positive feedbacks that could lead to the catastrophe that the 'cost-effective'

---

[*] Principal Advisor for Power and Energy, at the rank of Secretary to the Government of India. The views expressed are his personal views.

[1] Speech by the Rt Hon Gordon Brown MP, Chancellor of the Exchequer, at the Energy and Environment Ministerial Roundtable (G8 Environment and Energy Ministers' Meeting), London, 15 March 2005 (http://www.g7.utoronto.ca/environment/env_brown050315.htm).

stabilisation trajectory was designed to avoid in the first place. Thus if one was to impose a different set of value choices, one could argue, with equal legitimacy, that 5% or 10% of the global GDP would be required to reduce the risk of the catastrophic damage function to a benign 5-10% level. Now who decides which value choice or what probability of a catastrophic event occurring is acceptable. Who decides the value of an Indian life or, for that matter, how much more is life worth in the developed world?

Should we accept the 450-PPM (parts per million) trajectory that has even odds of setting in motion an irreversible melting of the Greenland ice sheet or would we rather be safer and accept the 2°C stabilisation bound that reduces the probability of such a risk to 20% or less. Remember, that the 450-PPM bound gives an extra 10-12 years of emissions growth before emissions need to decline sharply and the 550-PPM bound, more conveniently, doubles that period. An incremental approach that is politically acceptable is just politics and not a solution to the urgent problem of climate change.

The purpose of the foregoing discussion is not to establish the most appropriate stabilisation path but to demonstrate that climate change is primarily a political and socio-ecological issue. The socio-economic considerations, estimates of costs and benefits and who pays can be addressed only after political consensus emerges on ethical issues that would recognise the right of all human beings to a minimum development threshold. Such consensus would define national obligations towards global climate goals that are commensurate with each country's responsibility for the problem and their capacity to address global climate concerns while attaining the minimum development threshold.

Despite the urgency in developing a global climate compact, the climate debate is stuck because of unsustainable economic inequities. Populations are divided by wealth and other measures of well-being both among and within nations. A global compact that addresses both climate and inequality together is the only one that is likely to succeed. Further, such a compact must not attempt to distinguish investments in human development from adaptation activities. There are practical and conceptual problems with trying to determine the additionality of adaptation activities, and with trying to quantify incremental costs of adaptation over baseline costs of development.

Once one looks at the problem with the foregoing perspective, one recognises that the climate debate cannot be dealt with in isolation from the debate on globalisation, trade, intellectual property rights (IPRs), energy security and development. An environment and climate-friendly business community, investment and technology, though essential are not sufficient to drive climate change by themselves. If such market forces alone could address climate concerns then the world would not be in the current situation wherein the energy intensities of even the rich developed nations vary by a factor of two despite access to technology and funding and the presence of competitive and enlightened market players. Markets and businesses typically react to global political, social, ecological and developmental agendas – they do not and cannot provide the leadership to create consensus on such compacts. Such leadership lies squarely in the political domain and has, unfortunately, been missing.

The following eight inconvenient truths highlight the need for enlightened leadership from the developed world. In listing these inconvenient truths, the intention is not to cast any negative value judgement on any country or group of countries.

- First, climate change is already upon us because of natural drivers duly accelerated by anthropogenic activity.
- Second, the North has consumed the bulk of the global carbon budget while developing in an unconstrained world. There is little left for the South. Yet, the North's emissions will continue to rise till 2030 even under alternate benign scenarios. Both the IEA (International Energy Agency) and EIA (US Energy Information Administration) project that the world's fossil fuel dependence and the carbon intensity of its fuel mix will rise over the next 25 years.
- Third, a climate-constrained world imposes real limits to growth. Even if one takes the IPCC's modest 'B1' scenario for the developing world and plots it against the $2^{\circ}$ C or the 450-PPM scenario, it is clear that the growth in emissions from the South would hit a roadblock even as the South is fighting to meet the Millennium Development Goals and eradicate poverty. The South recognises this and, rightly, should not accept any uncompensated reduction in emissions that locks in poverty. The post-Kyoto negotiators would do well to recognise that this is not simply a bargaining position and should embrace the South's right to development at least up to a negotiated minimum threshold. Without such a realistic approach, the world will, together, fail to deliver global climate targets.

- Fourth, the IEA projects that the number of people dependent on biomass to meet up to 90% of their household energy needs will rise from 2.5 billion today to 2.7 billion by 2030. This will be a third of this future generation – almost mirroring the current share of have-nots. Climate negotiators from the developed world and the multilateral community must recognise that they are committed to delivering a threshold level of development to this third of the next generation. More importantly, this commitment takes precedence over all other global endeavours because delivering upon this commitment will give the world's have-nots a stake in its stability, thereby eliminating the most dominant source of conflict in the world.
- Fifth, current lifestyles and patterns of production and consumption in the North are simply unsustainable. The emissions from the North must peak soon – even yesterday may not be soon enough. Correcting this will entail huge costs and imposing them would be politically impossible without a binding global compact. Remember it has taken 15 years to introduce the concept of an 'emissions freeze' into the American political dialogue.
- Sixth, the North would need to fund even the no-regret mitigation options of the South since developmental outlays will always take precedence over mitigation in a resource-constrained South. And rightly so for growth and development also delivers the essential adaptive capacity as a by-product.
- Seventh, even if the North succeeds in bringing down its emissions to a level that is 80% below its 1990 level by 2050, the North's emissions would still be a multiple of its fair share under a per capita metric.
- And eight, nobody is forecasting a technological breakthrough by 2030. Our best bet is to make current low-carbon and energy-efficient technologies universally available and promote collaborative research.

Each of the foregoing inconvenient truths is a blazing turning point that suggests immediate action by the North. Yet the international community continues debating developing country commitments while ignoring unilateral initiatives by countries such as India that, although driven by national concerns of energy security, economic development and local environment, have, nevertheless, delivered global climate benefits. India presents a compelling example of the cost that the world's fourth-largest economy is paying due to inequity in the global energy/climate regime.

With 17% of the world's population, India consumes less than 4% of the global supply of fossil fuels. India's per capita commercial energy consumption is below 5% that of the US, below 36% that of China and below 22% of the world average. While India's energy mix is dominated by coal with a share of 35%, renewable energy follows closely with a share of 33%. Even though fossil fuels in India's energy mix will mirror the world average by 2030, renewable energy will remain the third most important energy source with a share of around 15%.

With 3.5 times the US population and more than twice the population of the EU, India has, since 2002, delivered more than twice their growth while consuming lower amounts of fossil fuels on an incremental basis in absolute terms – I repeat absolute terms and not in per capita terms. Compared to India, China's growth has been 25% higher but China's fossil fuel consumption has been over 9 times that of the EU, over 10 times that of the US and over 11 times that of India on an incremental basis since 2002 in absolute terms. In fact China's incremental fossil fuel consumption since 2002 is about 130% of India's total fossil fuel consumption.

India accounts for less than 4% of the global $CO_2$ emissions or about a fourth of the Chinese share. With an 8% plus GDP growth, Indian emissions are rising by 3.5% annually, whereas Chinese emissions are rising at over 9.5% per annum to support its 10% plus growth. More importantly, India's $CO_2$ intensity of GDP growth has been falling at the rate of 4.7% per annum.

India emits only 5-10% $CO_2$ compared to industrialised countries when it comes to putting food on the table, moving people or heating or cooling space. And per capita consumption of aluminium, cement and steel per unit of inhabited land area is also a fraction compared to industrialised countries. If India were to match the EU15's low consumption levels of these three construction materials, India's total emissions would rise to 2.7 times their current levels. Most importantly, recycling rates in India remain the world's highest.

The energy intensity of India's GDP growth, today, is half of what it used to be a generation ago and is the fifth lowest in the world. The Integrated Energy Policy of India, that I have recently crafted, details policy initiatives that will close even this gap. India has successfully decoupled economic growth from growth in energy consumption and has, in recent years, delivered 8% growth with just a 3.7% growth in primary energy consumption. While China adds 100 GW of power generation capacity annually, India's total installed capacity is only 150 GW. India's per capita energy consumption in 2031-32 will reach 15% of the current US level, 70% of the current world average and match that of China today!

India's achievement did not come without cost. In PPP terms, Indian taxes on energy and energy prices are the highest in the world. The paying Indians are being charged the highest tariffs for energy in the world in PPP terms. Indian lifestyles are far more sustainable and key energy-intensive industries have either achieved or are close to achieving world energy efficiency standards. India has recognised that energy efficiency and conservation provide the largest assured energy access and hence energy security to India.

Again, one must also look at the above Indian achievements from the perspective of 830 million Indians who, even today, live below the threshold of two dollars a day; or the perspective of over 700 million Indians who, even today, use some form of bio-mass for their predominant energy need, namely cooking; or the perspective of some 600 million Indians who, even today, live without electricity. While the world debates questions of global ethics, responsibility, costs and benefits of mitigation strategies, these fellow Indians, and the more vulnerable women and children among them, are busy combating local and indoor air pollution, unsafe drinking water, disease, infant and maternal mortality, illiteracy, gender bias, security of food and shelter - all key elements of a broad-based adaptive capacity; capacity that the multilateral community committed to deliver through the Millennium Development Goals and through Eradication of Poverty. Economic growth and the access it delivers offer the only hope that these fellow Indians have for their survival and empowerment.

The above analysis is not an attempt to assign blame, but simply to place facts on the table that show that there are differences among developing countries, just as there are differences among the developed countries. It is an attempt to show that while both India and China face common challenges, the tendency to talk of the two countries in the same tone is simply ill-researched. And finally, it is an attempt to show that some developing countries are unilaterally sharing the global climate burden well beyond their legitimate responsibility and capacity.

In conclusion one can say that climate is a global responsibility but equally, environmental space is a global common and all humans have the right to a threshold level of development. While we are all in this together, the South cannot address global climate concerns alone, just as the North cannot do it alone. The global climate policy-makers should ensure that available energy-efficient and climate-friendly technologies are put into limited public domain to avoid a carbon-intensive and business-as-usual Southern growth trajectory that repeats the mistakes of the North. The post-Kyoto regime must ensure that future energy research would be conducted under collaborative efforts with appropriate sharing of IPRs. The post-Kyoto regime must recognise that energy security is a global need and not just the right of some. Such a post-Kyoto regime would seek to curb unsustainable lifestyles wherever they exist, renew commitment to the Millennium Development Goals and eradication of poverty and ensure that the additional funds to achieve all of the foregoing would come from those capable of providing them.

# 5. Climate Change: A Threat to Development

*Anders Wijkman**

> *We basically have three choices – mitigation, adaptation and suffering. We are going to do some of each. The question is what the mix is going to be. The more mitigation we do, the less adaptation will be required, and the less suffering there will be.*
>
> John Holdren, President of the American Association for the Advancement of Science

The issue of climate change may seem remote to people living in low-income countries, compared to more immediate problems like poverty, disease and hunger. Yet we know that the consequences of climate change will affect their living conditions in a serious way.

The world's poorest nations bear little responsibility for the build-up of greenhouse gases (GHG) in the atmosphere, yet they are likely to suffer the brunt of the consequences of climate change. More extreme weather events, sea-level rise and water scarcity constitute serious threats to development efforts and to achieving the Millennium Development Goals in many regions of the world. The Fourth Assessment Report by the Intergovernmental Panel on Climate Change depicts the following consequences:

* Member of the European Parliament and Special Advisor to the Swedish Initiative on Risk Reduction and Adaptation.

**Africa**

- 75–250 million people across Africa could face severe water shortages by 2020.
- Agricultural production and access to food will be severely compromised due to lost agricultural land, shorter growing seasons and lower yields. In some countries, yields from rain-fed crops could be halved by 2020.
- Rising water temperatures will decrease fish stocks in large lakes, already depleted by over-fishing.

**Asia**

- Glacial melt from the Himalayas will increase flooding and avalanches, followed by reduced water supplies. Throughout Asia, the loss of fresh water could affect one billion people by the 2050s.
- In Central and South Asia, crop yields could fall by up to 30%, creating a high risk of hunger in several countries.
- Increased deaths and illness from diarrhoeal disease due to flooding and drought, as well as from cholera due to higher sea temperatures.
- Coastal areas, especially in mega-delta cities, will be at greater risk due to increased flooding both from the sea and rivers.

**Latin America**

- Shifting rainfall patterns and the loss of glaciers will significantly reduce water availability for human consumption, agriculture and power-generation.
- In dry areas, agricultural land will become salty and sandy, lowering crop yields and livestock productivity, thereby undermining food security.
- In tropical forests, higher temperatures and the loss of ground water will reduce biodiversity, affecting the livelihoods of indigenous communities.
- Rising sea levels will cause more flooding in low-lying areas, and warmer sea waters will diminish fish stocks.

### Small islands

- Rising sea levels and increased storm surges will threaten the homes and livelihoods of communities, forcing some to migrate permanently.
- Coastal erosion and coral bleaching will undermine incomes from fishing and tourism.
- Freshwater resources on small islands are likely to be seriously compromised, especially in the Pacific and the Caribbean.

### Climate change aggravates existing problems

While climate change will significantly increase the frequency and strength of extreme weather events, many regions of the world already have a long history of serious disaster-related problems. Both the frequency and scope of recorded disasters have risen markedly over time. A summary of natural disasters, compiled by Munich Re, world's second-biggest reinsurer point to a rapid increase over the past 50 years.[1] Figures from the International Red Cross[2] and elsewhere the corroborate this trend.

*Weather-related disasters* outnumber the *geo-physical* - by *nine to one* over the past decade! According to data from Munich Re, disasters caused by heavy storms and floods have increased six-fold since the 1950s. The Fourth Assessment Report by the Intergovernmental Panel on Climate Change (IPCC 2007) provides strong evidence that this increase will be exacerbated because of climate change.

Natural disasters have claimed the lives of more than two million people over the past 20 years, with an estimated 98% of the casualties occurring in low-income countries. For every reported death, there have

---

[1] The Munich Re website gives an overview on costs and losses of all major natural disasters since 1980 (http://www.munichre.com/en/ts/geo_risks/natcatservice/significant_natural_disasters/default.aspxMunich Re Annual Reports).

[2] "Preparedness for climate change: a study to assess the future impact of climatic changes upon the frequency and severity of disasters and the implications for humanitarian response and preparedness", study prepared by the International Federation of Red Cross and Red Crescent Societies in cooperation with the Netherlands Red Cross, 28th International Conference of the Red Cross and Red Crescent, Geneva, 2-6 December 2003.

been thousands of people seriously affected, often losing everything they have. Decades of development efforts are literally swept away.

Low-income countries are particularly vulnerable to natural disasters. Rich countries are far better placed to provide the infrastructure needed to reduce risk and vulnerability. Furthermore, in addition to the much greater capacity to mainstream risk reduction into development planning, the existence of insurance and re-insurance schemes provides important protection. In developing countries, vulnerable populations face the double jeopardy of being at much greater risk and having no capacity what so ever to transfer risk through insurance schemes.

## Disasters = Failed development

Natural disasters occur when communities are exposed to potentially hazardous events without being able to absorb the impact. While it is common to talk about natural disasters, both vulnerability and hazard are conditioned by human activities. Reducing the effects of disasters means tackling the development challenges that lead to the accumulation of hazard and vulnerability that preceded the disaster. As underlined by a UNDP (2004) report "Natural disasters destroy development gains, but development processes themselves play a role in driving disaster risk."

When looking for the causes behind the rapid increase in natural disasters, we are confronted with a complex reality. In addition to climate change, a host of factors are of specific importance:

- *Poverty.* Extensive poverty inevitably leads to *increased vulnerability* in the face of various natural hazards. People are forced to settle down or remain in highly risk-prone areas.
- *Population growth.* Demographics is an important element in explaining the increase in the number of natural disasters and disaster victims. While fertility rates have decreased in most countries over the past decades, the growth in numbers is bigger than ever before. A large proportion of the new inhabitants end up in risk-prone areas.
- *Rapid urbanisation.* While the global population has doubled over the past 40 years, the number of people living in urban areas has increased five-fold. Most of the new residents in urban environments end up in unplanned slums where no mitigation measures have been

taken against natural hazards such as earthquakes, flooding and tropical storms.

- *Environment degradation.* When forests are indiscriminately logged, soils eroded and wetlands are diked, future natural disasters are in the making. The irony is that while disasters are triggered by natural phenomena, a healthy natural environment is often the best possible protection against storms, heavy rains or droughts turning into disasters.
- *Disaster prevention overlooked.* Although disasters are no longer viewed as extreme events caused by natural forces alone, disaster prevention and risk reduction are often overlooked. This comment goes both for national governments and development agencies. Out of more than 60 poverty-reduction strategies prepared to date, less than a fifth have incorporated aspects of hazard risk management.

**Risk reduction a priority in the North...**

Western countries spend billions of dollars at home in reducing the risks associated with floods, heavy storms and earthquakes. Many new initiatives have been launched to assess the increased risk panorama because of climate change and to suggest investments for risk reduction.

The European Commission's recently launched a Green Paper on "Adapting to Climate Change in Europe - Options for EU Action" (European Commission, 2007), which underlines the scale of the challenge facing Europe. The document warns that, unless there is advance planning, European countries will be left to respond "to increasingly frequent crises and disasters which will prove much more costly and also threaten Europe's economic and social systems and its security".

The Green Paper identifies a host of problems, like health risks and social problems, including the threat to employment and decent living and housing conditions. Moreover, climate change will affect crop yields and livestock management, putting farm incomes at risk in some regions. The document predicts that climate change will specifically affect industries like construction and tourism: "Major infrastructure such as bridges, ports and highways have lifetimes of 80-100 years, so today's investments must take full account of the conditions projected for the end of the century."

If all this is true for countries in Europe, the situation is many times more serious in many low-income countries!

## ...but mostly overlooked in the South

Low-income countries are by far the most vulnerable to disasters and the increased risks of climate change. As explained above, current development patterns often result in increasing vulnerability of local communities in the face of natural disasters.

In spite of the often-devastating impacts that natural hazards can have, most policy-makers have been hesitating to commit significant funds to risk reduction. Yet, there is ample evidence that mitigation pays.

Many studies have demonstrated the cost-effectiveness of prevention and risk reduction, both in terms of lives saved and 'disaster-related' economic losses. On a general level, it is estimated that €1 invested in disaster-risk reduction equals €5-10 invested in dealing with the disaster once occurred. The cost-effectiveness of early implementation of adaptation measures is particularly high for sectors like infrastructure. Furthermore, many ongoing development activities - like the destruction of coastal mangroves and coral reefs and/or human settlements in risk-prone areas - may irreversibly constrain future adaptation efforts.

Although its long-term benefits may seem obvious, there are several reasons why risk reduction and adaptation are often overlooked:

- Short timescales often dominate development planning, causing a general neglect of disaster risks in decision-making. More specifically - and as explained in a recent OECD report[3] - there is also often a mismatch between the time and space scales of climate change projections and the information needed by development planners. The main sensitivity to climate change is usually at a local scale, for which credible data are often lacking.
- Mainstreaming of climate change risks often proves difficult because of the pressing needs that governments feel in relation to more immediate issues, like poverty.
- It is easier to raise funding for high-profile humanitarian relief operations than for prevention. As Kofi Annan once expressed it: "The disaster that did not happen will create no headlines."

---

[3] S. Gigli and S. Agrawala, *Stocktaking of Progress on Integrating Adaptation to Climate Change into Development Co-operation Activities*, COM/ENV/EPOC/DCD/DAC(2007)1/FINAL, OECD, Paris, 2007.

In addition, developing country governments as well as aid agencies seldom command the overall skills and resources to effectively address disaster risks. The current challenge is to raise adaptation to the top of the political agenda - both in terms of policy priorities and in terms of funding - despite these inherent obstacles.

**Costs of risk reduction staggering**

Millions of poor people who are already experiencing the effects of climate change are doing their best to find ways to cope with it. The ADAPTIVE research project[4] found that people in Africa were responding to less frequent rains by planting more drought-resistant crops, eating wild fruits, collecting wild seeds, selling their animals, seeking paid jobs in towns, etc.

But there are clear limits to how far poor people can adapt without support from the outside. Many people lack viable opportunities to diversify their livelihoods and also lack access to reliable climate information that would help them to plan better.

The most obvious place to provide leeway for adaptation and risk reduction measures would be in Poverty Reduction Strategies (PRSs), providing the framework for the low-income countries' long-term development planning. However, most of the PRSs so far adopted have made limited efforts to incorporate aspects of hazard risk management. There are exceptions, like Honduras, Nicaragua, Vietnam, Bangladesh and Mozambique. But for most of the PRSs, disaster risk reduction is not a priority, and in most cases is not even mentioned.

Cost estimates for adaptation to climate change differ widely. According to the Stern report, the additional costs of making new infrastructure and buildings resilient to climate change in OECD countries could be between $15-150 billion per annum. Taking into account only the costs required to 'climate-proof' investments, the World Bank (2006) has estimated that the annual adaptation costs in developing countries could range anywhere between $10 billion and $40 billion. In a recent report by OXFAM (2007), it is suggested that adaptation in developing countries will cost at least $50 billion each year. Evidently, there is a great need to gain a more precise picture of the costs for adaptation.

---

[4] See http://manage.gov.in/PAR/about.htm.

## The international community: Insufficient response so far

The International Strategy for Disaster Reduction (ISDR) developed by the United Nations is one attempt to help reduce vulnerability among the billions of people who live in disaster-prone areas. The ISDR has seen increased momentum after the World Conference on Disaster Reduction, held in Kobe in January 2005. The Conference adopted the Hyogo Framework for Action (HFA), with its five priority areas, with a strong focus on action at the national level. After the follow-up *Global Platform on Disaster Reduction* in 2007, ISDR has been reformed into a system to support the wide compliance with and implementation of the HFA.

Several humanitarian organisations have also been pro-active in calling for more attention to risk reduction and disaster prevention. One prominent example is PROVENTION - a global coalition of international organisations, governments, the private sector, NGOs and academic institutions dedicated to increasing the safety of vulnerable communities and reducing the impacts of disasters in developing countries. PROVENTION acts somewhat as a 'think tank' for the international system, in particular on tools for mainstreaming risk reduction and adaptation into development planning.

The Framework Convention on Climate Change (FCCC) recognises the right of poor countries to receive support in adaptation to climate change (e.g. Articles 4.8 and 4.9). But action has been slow. Three different funds of a voluntary nature have been established to help support developing countries address different aspects of climate change:

- *Least Developed Countries (LDC) Fund* - whose main objective is to support LDCs to carry out assessments of their National Adaptation Plans of Action (NAPAs); so far only a few countries have made commitments to this fund and only a handful of NAPAs have been undertaken.
- *Special Climate Change Fund* will assist developing countries - not just the LDCs - to deal with climate change, including adaptation as well as mitigation measures. So far an estimated $500 million have been committed.
- *Adaptation Fund,* which will consist of the proceeds of a special adaptation levy placed on transactions under the Clean Development Mechanism. So far the levy has mobilised very limited funds.

In addition, the Global Environment Facility (GEF) has established a special fund – *Strategic Priority on Adaptation* – as a three-year initiative to pilot capacity-building adaptation measures in developing countries. Funding so far amounts to $50 million.

Yet another international mechanism should be mentioned – the *Global Facility for Disaster Reduction and Recovery (GFDRR)* – launched by the World Bank in 2006. The role of the GFDRR is to offer technical assistance at country level, primarily through country-risk assessments. The facility does not engage in risk-reduction investments, however.

It is also worth mentioning that the European Commission currently is in the process of drafting a *Communication on Disaster Risk Reduction (DRR)*, with the purpose of linking adaptation with the broader disaster prevention agenda. This is a welcome step towards a more coherent approach.

Although progress has been slow and governments – both in the North and the South – are only starting to pay serious attention, many experts now see a momentum emerging, aiming at expanding the political space dedicated by governments to disaster risk reduction . But the overall picture of risk reduction and adaptation still remains fragmented and the funding required to effectively address risk reduction in low-income countries is simply not there.

Compared to the estimates provided both by the World Bank (2006) and OXFAM (2007) with regard to adaptation and 'climate-proofing', the three voluntary funds within the UNFCCC as well as the initiatives by the GEF and the World Bank only represent a very modest start. New and additional resources have to be mobilised – either through dedicated funding streams for adaptation, like a special global levy on air travel or part of the revenues from future auctioning of emission permits, through increased ODA (official development assistance) or through some kind of market-based mechanisms (by creating a market in 'adaptation credits' or 'vouchers').

## Swedish initiative in the making

Tackling climate change will require a high degree of cooperation at the international level. The evidence so far of such cooperation is limited. Industrialised countries have only just started curbing their own emissions – in fact, emissions continue to increase in most Annex I-countries – and efforts to assist developing countries both in mitigation and adaptation have been extremely limited.

If we are genuinely interested in a post-2012 agreement, the situation must change dramatically. Industrialised countries have to recognise their responsibility for historical emissions and help developing countries to reduce disaster risks as well as curb emissions.

Why otherwise would developing countries agree to be party to an ambitious international agreement on emission reductions?

A *crash programme* is needed at international level to address the problems of adaptation and risk reduction. The main focus should be on *weather-related disasters* and impacts related to climate change. However, it is not possible to isolate the impacts from climate change from natural variations with regard to extreme weather events. Therefore a comprehensive approach to risk reduction should be pursued, one that recognises all major disaster risks.

The risk reduction and adaptation initiatives that have been launched recently – like International Strategy for Disaster Reduction (ISDR), a global platform for disaster risk reduction, the World Bank Global Facility, UNDP GRIP (Global Risk Identification Programme) initiative on disaster risk mapping and the efforts of PROVENTION on vulnerability assessments – merit attention and support. But what has been done is far from sufficient. What is badly needed are systematic efforts that pull together the many scattered initiatives undertaken so far.

To respond to the needs, the Swedish Government recently launched an initiative that hopefully will make a difference. An *International Commission on Risk Reduction and Adaptation* will be established under the chairmanship of the aid minister, Gunilla Carlsson. The Commission's main task is *to explore and promote effective ways to integrate risk reduction and adaptation to climate change into development and poverty reduction plans in developing countries and to ensure that future investments in ODA take full account of climate stresses and increased disaster risks.*

The work shall concentrate on the following main objectives:

***First***, identify and analyse the incentives as well as barriers for developing countries to undertake risk-reduction and climate-proofing measures in their development efforts. The main focus shall be on weather-related disasters and impacts on development related to climate change. A comprehensive approach to risk reduction, recognising all major disaster risks shall be pursued.

***Second,*** consider how to best combine long-term efforts for climate change mitigation and the urgent need to support adaptation efforts in developing countries as a result of climate change.

***Third,*** identify directions for international development cooperation in the field of adaptation and risk reduction, with a focus on leadership required in development assistance to ensure integration of long-term objectives such as risk reduction, adaptation and resilience in policy-making.

***Fourth,*** consider how to achieve policy coherence by integrating concerns for climate change in wider development efforts and elaborate a methodological approach for an Integrated Analysis for Climate-Proof Development.

The Commission shall also give specific attention to the following:

- Help increase awareness about the need to mainstream risk reduction and adaptation into Poverty Reduction Strategies and ensure coherence in development strategies;
- Explore and propose new methodologies for integrating climate change aspects into development strategies;
- Assess the role of ecosystems in disaster prevention and devise strategies to strengthen their capacity to meet climate change;
- Give special attention to the hazards experienced by an estimated one billion slum dwellers and identify people-centred solutions to reducing their vulnerability;
- Give priority to slowly developing disasters, such as prolonged droughts and chronic instability stemming from water scarcity;
- Assess if and in that case how risk management mechanisms within the insurance industry can be used in informing about risk reduction and adaptation measures in developing countries.

The *raison d´etre* for an initiative like this are obvious. The Millennium Development Goals are seriously at risk unless disaster risks can be significantly reduced. Moreover, the possibilities to strike a post-2012 climate agreement very much depend on a much greater readiness by industrialised countries to meet their historical responsibilities and help developing countries cope with the consequences of climate change. I sincerely hope the Swedish initiative will be able to add value and help strengthen the worldwide efforts to mainstream risk reduction and adaptation into development and poverty reduction plans.

## References

European Commission (2007), Adapting to Climate Change in Europe – Options for EU action, Green Paper from the Commission, COM(2007) 354.

Gigli, S. and S. Agrawala (2007), *Stocktaking of Progress on Integrating Adaptation to Climate Change into Development Co-operation Activities*, COM/ENV/EPOC/DCD/DAC(2007)1/FINAL, OECD, Paris.

IPCC (2007), Climate Change 2007 – Impacts, Adaptation and Vulnerability, Working Group II Report, IPCC Fourth Assessment Report.

Oxfam International (2007), *Adapting to Climate Change: What's Needed in Poor Countries, and Who Should Pay*, Oxfam Briefing Paper 104, Oxfam International Secretariat, Oxford.

Stern, N. (2007), *The Economics of Climate Change*, Cambridge: Cambridge University Press.

UNDP (2004), "Reducing Disaster Risk: A Challenge for Development", Report by United Nations Development Programme, New York.

World Bank (2006), *Clean Energy And Development: Towards An Investment Framework*, World Bank, 5 April 2006.

# 6. THE BUSINESS CASE OF CLIMATE CHANGE

*LARS G. JOSEFSSON*[*]

## 1. How companies are affected by climate change

Gudrun is an ancient Nordic name meaning prophetess. But ever since 8 January 2005, the name has taken on another, more sinister meaning.

It began with a storm warning out in the North Atlantic. Soon the first reports of flooding and damage in Scotland and northern England were coming in. A freighter off the coast of Denmark foundered after its crew had been safely rescued.

Gudrun was a storm. With wind speeds of over 80 mph, she reached hurricane strength at times. This can seem modest compared to the hurricanes spawned in the Caribbean basin, but in northern Europe it is unusual - and therefore deadly. When Gudrun took aim on southern Sweden, she chanced upon an unprepared landscape. Within the space of just a few days 250 million trees were felled. Roofing tiles flew, nuclear power stations suffered emergency shutdowns and the first reports of fatalities soon arrived.

No one can say with certainty that Gudrun was a result of global warming. What we do know however, is that climate change will bring about unusual meteorological phenomena in the future, and one such phenomenon is severe storms.

And how does this pose a risk for business? The answer is quite simple. All companies with sensitive infrastructures spread out over wide areas can be hurt badly by storms.

---

[*] CEO, Vattenfall.

In the specific case of Gudrun, 20,000 kilometres of power lines were wiped out within the course of a few hours. Seen from the electricity companies' perspective, it naturally involved great costs for repairs and new investments – the total bill is yet to be presented.

But this perspective is too narrow. A natural catastrophe cannot be viewed solely from the perspective of the business world. The power lines that were destroyed supplied 630,000 households. Of these, 12,000 customers had to wait 20 days before power was restored. Others were forced to hold out for months.

Gudrun illustrates how the climate threat is a danger to our entire habitat, and as such a danger to people and companies alike. In other parts of the world, where a lack of drinking water can be the most serious result of global warming, large areas risk becoming uninhabitable. How can companies flourish in places where no one can survive? Who wants to work where noone wants to live?

Therefore we can draw the simple conclusion that humanity and business are mutually dependent upon one and other. A threat to humanity is just as much a threat to business. And this brings us to another question: why should companies get involved in the climate debate?

Since 2005, Vattenfall has pursued a dynamic climate initiative. It is a commitment that reaches above and beyond those formal demands placed on a viable business enterprise that authorities and other interested parties usually direct towards an energy supplier. Our commitment started with a concrete, business-critical issue, namely what position should we adopt regarding the coal-fired power stations that were a significant part of the acquisitions we made in Germany 2001–03. It is well known that carbon dioxide is an inescapable bi-product of coal combustion. This involved a new kind of environmental outcome for Vattenfall, which until then had primarily dealt with safety issues and environmental consequences regarding hydro-electric and nuclear power. The company management – myself included – realised that we needed to learn more about the complexities of the climate problem. Much more. Otherwise we could not have carried through the acquisition with sufficient background knowledge. Thus far was the climate threat a threat to our company.

During the past year climate change has come to be perhaps the foremost globally debated issue. This was not the case in 2003, and therefore I was personally surprised and alarmed when I realised just how serious the threat was and how little had been done to counter it. At that precise moment the basis for our involvement changed. From having been about coal-fired power stations and Vattenfall, the issue changed to one concerning our responsibility as a large energy company to contribute towards solving a problem shared by all of humanity.

The answer to the question of "Why should companies get involved in climate issues?" is therefore ... because we must. As companies we are also part of society. As a private individual, I am part of society.

This may sound a bit obvious. The problem is that the role of business in the climate debate has not been obvious. We are seen sometimes - by politicians, pundits and environmental organizations - as being a narrow special-interest group with a hidden agenda. Before any of our suggestions are ever considered our intentions are always considered first. And in some ways I can understand this. Traditionally industry has taken a defensive position on the environment. Whenever new restrictions find favour with legislators, companies either adapt or move on somewhere else. Seldom has the initiative for restrictions been taken by industry itself.

But the climate issue is a genuine global challenge. If one of us loses, we all lose. It is therefore extremely important for me to emphasize the mutual dependency which is the mainstay of our commitment. Not just to be seen as credible but also because the ideas and models in our suggestions build on the whole world's - every nation, every community and every individual - reaching agreement on what must be done.

## 2. How companies are affected by climate policy

The discussion on how climate policy affects companies can be divided into two parts. Firstly, how it ought to affect us; i.e. the ideal situation, and secondly, what risks arise for industry if the ideal situation cannot be realised.

In connection with the World Economic Forum in Davos in 2006, Vattenfall introduced 'Curbing Climate Change' - our model for how all the nations of the world can cooperate to reduce emissions to an acceptable level. It builds on the following main elements:

- Up until 2100, world nations should not discharge more than a total of 1,600 billion tonnes of carbon dioxide. The figure is an estimate - in the final stage this level may be adjusted up or down. But if we manage to restrict emissions more or less to this level we will avoid concentrations of carbon dioxide in the atmosphere being altogether too high, which would directly affect global warming.

- Even though 1,600 billion tonnes does sound like a lot, it represents in fact a gigantic reduction. It requires our reducing today's emissions by around 80%.
- How will the burden of reductions be shared among the nations of the world? For a start, we must all be part of the same system. The situation today, where the EU has taken it upon itself to abide by the Kyoto Protocol, is unfair and untenable in the long term.
- The system in Vattenfall's model is based on every country being allotted a limit to its emissions. How these limits are made use of is a matter for individual governments. At the end of the chain, every company is allotted a ceiling for its emissions – let us call it emissions rights – which management can use for their business or sell on.
- How will allocation be accomplished? Vattenfall has chosen a simple method: You pay what you can afford. Calculated per unit of GDP, developing countries may emit more carbon dioxide compared to richer nations. The gives developing countries the opportunity for continued development, and ensures at the same time that they must contribute something. As prosperity increases, so too does the requirement for restrictions.
- The very poorest of nations would be subjected to no restrictions at all. Every affluent country in the world must contribute from the very beginning, but there is an upper limit to how quickly restrictions are applied, to avoid creating economic crises in individual countries.
- If market forces are allowed free rein, global prices for carbon dioxide emissions would soon be established.

And how would industry be affected by such a model? Because states – and not companies – constitute the basis for the distribution of emissions reductions, companies will be affected differently depending on where their business is conducted. Yet this difference will be lower compared to today's situation where European companies are subjected to restrictions while companies in other parts of the world are not compelled to do anything. In Vattenfall's model, all of the world's companies – via their home countries – are part of the same system and contribute over the long term to the essential reductions of emissions.

Since Vattenfall's model was introduced, as just one of many suggestions for discussion, the global debate has evolved. An American proposal has been presented to begin discussions on a coming climate regimen. Additionally, the EU, during the German Presidency led by Angela Merkel, has agreed on new objectives for reductions aimed at 2020.

I was pleased to note that several of Vattenfall's ideas were raised in the discussion, among others the need for a global effort (everyone must be on board) and the necessity for market solutions and a worldwide carbon dioxide price.

But it remains unclear what mechanisms will be constructed for sharing the emissions burden. Or what time frame will be applied. If we end up too far from the ideal situation, which in my world is the same thing as Vattenfall's model, my fear is that several risks to business can arise. Here are the most manifest:

- *Uncertain business conditions.* For industry to flourish, the prerequisite has always been reliable, secure and transparent business conditions. Whenever a sector is threatened by regulation, investments are put on hold until the new rules of the game are known. This is the kind of uncertainty factor that can result from climate policies. Uncertainty spreads in many ways. Consider emission rights in Europe. Is it only our continent that must tighten its belt? Which sectors will be included and which will be exempted? Will global transportation continue to be excluded from carbon dioxide charges? In today's state of affairs, emissions from production plants can be 'exported' from Europe (production takes place where no regulations apply) only to be 'imported' again in the form of finished products. Even within Europe, emission rights cause consternation. In Sweden, for example, significant investments in natural gas-fired heat and power stations were called into question when the allocation of emissions rights was taken into account. Thus a politically-led, relatively short-term allocation became a determining factor in long-term investment decisions. Another aspect, which is peculiar to the energy sector in which Vattenfall is active, concerns the regulation of other forms of energy. In many parts of the world, nuclear power is recognised as a competitive, carbon dioxide-free alternative. But in a number of countries there is considerable uncertainty at the political level as to whether or not nuclear power is an energy source that can be expanded and improved. This uncertainty naturally results in no investments being made.

My message to politicians is: do not exacerbate this uncertainty with short-term, unclear decisions. The clearer the rules of the game are, the easier the transformation to a low-emissions economy will be.

- *Risk of unnecessary value destruction.* Whenever the volume of the climate debate is turned up, the risk increases for rash decisions on how emissions reductions should be achieved. It is easy to become fixated on those plants and machines that emit the largest quantities of carbon dioxide. Certainly, all industrial enterprises should in principal strive to achieve zero emissions in the long term, but we also need to work out where we should begin. It would be reasonable to seek out the easiest and cheapest solutions first. And this is also the idea behind the market solutions to reductions in place. If we can arrive at a global system based on market principles, then the transformation to a low emissions society will occur naturally. In the worst case, its opposite would be a mad scramble where desperate, dirigist powers-that-be prohibit certain types of business activity. The latter scenario does not just result in great costs to society. It may also be ineffective.
- *Energy policy as foreign policy.* There was a time when countries with abundant energy resources could attract direct investments by offering low electricity prices. Today there are few who can afford to use electricity as a come-on. Energy has become far too valuable. Sometimes so valuable that it can be used as a foreign policy weapon. Such a situation risks running off the rails with war as the ultimate consequence, which would entail both great suffering and huge costs for society. However, this risk would be much lower if all nations part of a long-term system.

## 3. What can companies do about climate change?

In opening, I declared my views regarding the role of companies as community citizens, and the importance of understanding the mutual dependency between industry and people. With this approach comes responsibility. Merely noting social responsibility is not enough. Business executives who want to take their role seriously must be prepared to act – not least when it comes to global warming, which is the great challenge of our time. Which raises the next question: what can individual companies do?

At this point it would be easy to be oh-so-wise and serve up edifying pieces of advice. Instead of telling others what they ought to be doing, I would rather provide a few examples from Vattenfall's – and my – journey to becoming committed contributors to the climate debate.

Before Vattenfall began to ponder its acquisition of the German businesses, i.e. before 2003, I must honestly admit that I was not well-read on the subject. My knowledge of the effect that carbon dioxide emissions had on climate was at best that of a inquiring engineer's curiosity about current events. I kept abreast from a distance; stayed up-to-date. But I had no real insight – it had never occurred to me just how enormous and essential reductions were.

As I mentioned, my road to understanding began when I was defending my company's interests. I wanted to examine what it would entail to own coal-fired power stations in an age when more and more raised voices were saying that climate change was accelerating. I soon realised that this issue concerned much more than just Vattenfall's acquisitions in Germany. It was about the future of us all – and I bemoaned the absence of industry's voice in the debate.

That insight was the first step. Ever since then, we have acted. The model Curbing Climate Change gave us a theoretical foundation. It was followed by a book, "The Future in Our Hands", in which I explain the model and why Vattenfall became involved in the climate issue. The book was published in five languages, and was distributed to all of our employees. The next big campaign was Climate Map – a survey showing how emissions can be dramatically reduced by quite simple means. Using concrete examples we show how all of industry, divided into six large sectors, can effect reductions by the year 2030. The goal for that year is to emit no more than 31 billion tonnes of carbon dioxide equivalents globally. In this way we can prevent the concentration of greenhouse gases in the atmosphere from exceeding 450 ppm. If we do nothing at all, emissions risk exceeding 58 billion tonnes, which is untenable.

Models and books are all well and good. The most effective means of communication is nevertheless meetings between people. A large part of my time as CEO has therefore been dedicated to taking part in various forums, among others the G8 Climate Change Roundtable. At the end of 2006, I was appointed Advisor for Climate Issues to the German

Chancellor, Angela Merkel. In this capacity I have had the privilege of contributing knowledge and putting forward arguments in a forum I otherwise most likely would not have had access to.

Vattenfall is also a driving force in its own forum, namely the previously-mentioned '3C - Combat Climate Change' a business leaders' initiative'. The number of companies rallying to the call grows continually and at the time of writing is more than 40.

To further emphasise the long-term nature and strength of our commitment, a special department responsible for climate policy at Vattenfall was established at the beginning of 2007.

In conclusion, I would like to underline that it is not the responsibility of companies to formulate coming climate regulations. This falls to the democratically elected representatives. Our role is to contribute knowledge, form opinions and support politicians in their efforts. We have suggested models and shown potential ways ahead, and now we look forward to a resolution that aims at a global system to succeed the Kyoto Protocol. Thanks to our quest for knowledge, I know this to be possible. All we need is the will.

## APPENDIX
## Climate Map - A summary

Together with McKinsey, Vattenfall has studied the potential for radically reducing the emissions of greenhouse gases by the global economy over the next 25 years.

If nothing is done, the total annual emissions will increase from 40 billion tonnes of carbon dioxide equivalents in 2002, to 58 billion tonnes in 2030. The emissions level in 1990 was, calculated in the same way, around 35 billion tonnes. For a development that restricts the total greenhouse effect to 2°C to be possible with a reasonable degree of certainty, the persistent content of greenhouse gases in the atmosphere must be limited to 450 ppm ($\pm$ 50 ppm). By 2030 total emissions of greenhouse gases must be restricted to 31 billion tonnes, i.e. a reduction in the order of 27 billion tonnes must be achieved by that year.

Vattenfall's survey shows this to be eminently possible. More than two-thirds of the measures for 2030 can be effected with solutions already available today.

How can the change be made? We must set a price for emissions and in this way utilise market forces to effect the change. The nations of the world must agree to binding emissions restrictions. For this to be possible, restrictions must be drawn up such that they do not present an obstacle to development and do not cause economic shock for any individual nation, and at the same time their effect on international competitiveness must be moderate and acceptable to all concerned.

A summary of the most important steps in such a process is set out below:

- Firstly, there must be a mutual ambition to reach a mutual goal, which ultimately must be expressed as a temperature. Given the knowledge we have today, it is reasonable to attempt to limit global warming measured as an increase in the mean temperature measured at the Earth's surface to 2°C in relation to the pre-industrial level. This means that concentrations of greenhouse gases in the atmosphere must be limited to 450 ppm ($\pm$ 50 ppm).
- Secondly, a framework must be established that makes possible the creation of value from emissions reductions. Value creation is possible when the credibility of emissions pricing reaches a level at which revenue and cost flows can be capitalised. Mechanisms for trading emissions must be maintained and developed further.
- Thirdly, mutual efforts must be made leading to ever-better solutions being developed, and their being made available globally. This will involve demands being placed on products, systems for marking and the dissemination of information, investments in the propagation of knowledge and establishing proficiencies. It will also involve efforts that increase and accelerate technical exchange by supporting the development of key technologies and measures that speed up their market introduction.
- Fourthly, it will involve finding ways for the mutual assumption of responsibility regarding the necessary adjustments to the consequences of climate change.

The climate issue is about lasting welfare development and global stability. The give-and-take necessary for establishing global understanding must occur on this basis. The foundation must be a common acceptance of responsibility and mutual commitment. The survey of potential measures carried out by Vattenfall shows clearly that it is not possible to counter the climate threat by efforts in certain regions or business sectors; the entire global economy must make the changes.

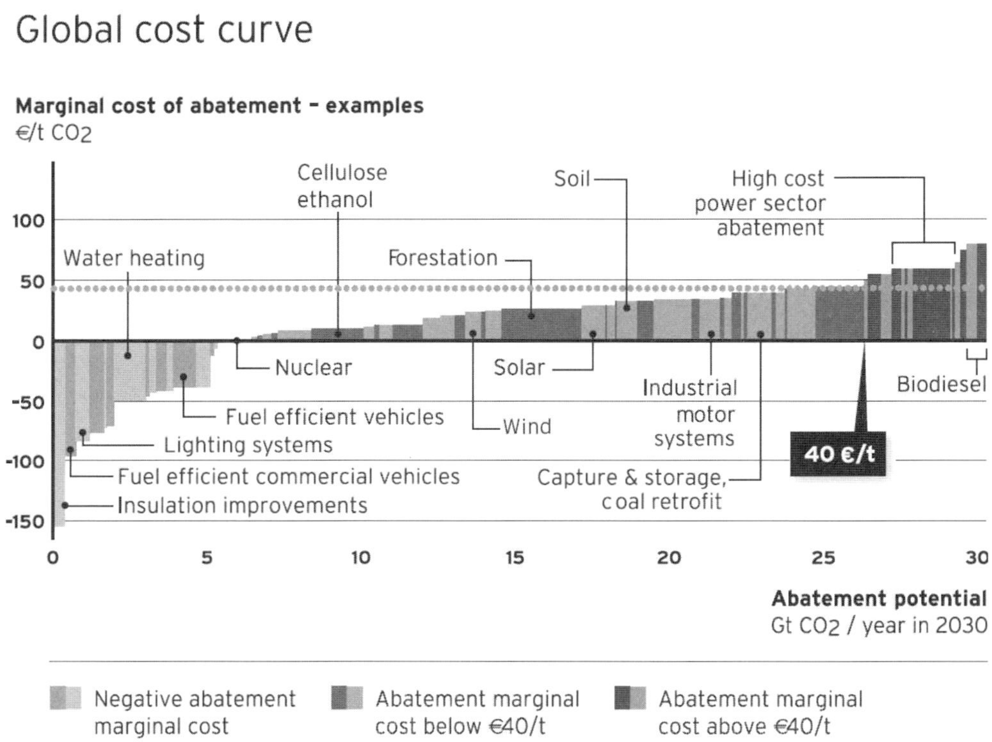

# Part II
# Research Contributions

# 7. ADAPTATION AS A STRATEGIC ISSUE FOR THE CLIMATE CHANGE NEGOTIATIONS

*SIVAN KARTHA**

## 1. Introduction

The tragic irony of climate change is that those who are the least responsible are the most vulnerable. It is the poor in developing countries who will bear the brunt of its impacts. Industrialised countries, including the member states of the European Union, bear a large degree of responsibility for these impacts and are thus morally obliged to help shoulder the costs of adaptation. But it is also in their self-interest to do so. In this increasingly globalised world, the effects of insufficient adaptation in developing countries will be felt in industrialised countries, for example through a decline in world trade activity, an increase in the spread of disease, escalating disaster-relief expenditures and increased migration flows (IPCC, 2007). To the extent that countries might ultimately be held

* Sivan Kartha is Senior Research Fellow, Stockholm Environment Institute (SEI) - US and a member of the Clipore Research Team. The chapter is a slightly revised and updated version of "Adaptation as a Strategic Issue in the Climate Negotiations" by Sivan Kartha, Preety Bhandari, Louise van Schaik, Deborah Cornland and Bo Kjellén, published as ECP Report No. 3 in November 2006. At the time of the publication of the ECP report, Preety Bhandari was a researcher at The Energy & Resources Institute (TERI), New Delhi, India and a member of the Clipore Research Team; Louise van Schaik, is Research Fellow at the Netherlands Institute of International Relations 'Clingendael' in The Hague, the Netherlands; Deborah Cornland is the Clipore Project Director and Bo Kjellén is Senior Research Fellow of the Stockholm Environment Institute (SEI) and ECP Co-chair. The many comments the authors have received on various drafts of the report are gratefully acknowledged.

legally responsible for causing climate change in other countries, ignoring the issue might lead to legitimate claims for compensation in the future. Moreover, a global or near-global climate mitigation regime would be difficult to negotiate without an effective framework for addressing adaptation. This is because meaningful support for adaptation is likely to be a condition for developing countries to participate in mitigation aspects of any future climate regime in a meaningful way.

This chapter examines the nature of adaptation and draws some lessons regarding the characteristics of a climate regime that would enable sufficient and timely adaptation in developing countries. This is a topic that has risen in prominence on the agenda of the international climate change negotiations under the auspices of the United Nations Framework Convention on Climate Change (UNFCCC) and other international forums as well.

The 'Bali Roadmap,' agreed at COP 13 in Bali, Indonesia, has further fixed adaptation as one of the four pillars of an effective global climate effort. Establishing a new body known as the 'Ad Hoc Working Group on Long-term Cooperative Action under the Convention', the goal is an agreement covering adaptation, mitigation, technology and provision of financial resources, in time for adoption at COP 15 in 2009.

## 2. Adaptation concepts

The close link between climate change and sustainable development is now universally acknowledged. Human-induced climate change will adversely affect agricultural and hydrological systems, forests, fisheries, health and economic infrastructure, and other natural and socioeconomic resources. The impacts will exacerbate existing conditions of poverty, malnutrition and illness, thereby posing a threat to the achievement of development objectives, including the Millennium Development Goals (MDGs) (see also chapter by Sethi elsewhere in this volume).

There is a growing understanding that, to be effective, adaptation efforts must address vulnerability, that is, build resilience to stresses on economic, social, political and environmental systems (Smit & Pilifosova, 2001; Turner et al., 2003) A comprehensive definition of adaptation comprises two complementary elements: implementing *adaptive responses* and enhancing *adaptive capacity*. Adaptive responses refer to measures adopted in response to existing or anticipated climate impacts. Adaptive capacity refers to the capacity to identify, assess, modify and implement effective adaptive responses without compromising future adaptive capacity (see also chapter by Wijkman elsewhere in this volume).

*Adaptive responses* should not be interpreted too narrowly. Localised, technology-based and infrastructure-focused interventions that anticipate a specific climate impact and are targeted at a specific sector are too limited for a number of reasons (Klein et al., 2007; Patt et al., 2005; Burton & van Aalst, 2004). First, the reliance on climate change projections makes the long-term effectiveness of specific, localised adaptive responses subject to great uncertainty. Second, individual adaptive responses tend to be partial - often short-term - solutions. Third, when the definition of adaptation is limited to responses that are specific to climate change, it neglects the fact that vulnerability to climate change rarely occurs in isolation, but rather as a syndrome accompanied by other types of vulnerability. For example, it may help to provide a rural household that grows a particular subsistence crop with a more drought-resistant variety, but a robust and comprehensive strategy would seek to improve food security generally through a set of coordinated measures that includes, say, agricultural extension, crop diversification, integrated pest management practices and rainwater harvesting.

The concept of building *adaptive capacity* is yet more fundamental and far-reaching (Smith et al., 2003). It recognises that it is not possible to anticipate the exact impacts of climate change and put in place precise defensive mechanisms to deal with them and the multiple other stresses that vulnerable communities face. It recognises that these combined stresses create a syndrome of vulnerability that can best be addressed through fundamental investments in building resilience. For example, the rural household is most capable of successfully implementing adaptive responses (i.e. the set of food security measures mentioned above) if it has adaptive capacity in the form of, say, access to investment capital through local financial institutions, close integration into intact social networks, an open channel for conveying concerns and priorities to local decision-makers, and a literate family member.

The report entitled *Poverty and Climate Change: Reducing the Vulnerability of the Poor through Adaptation*, issued by eleven bilateral and multilateral development agencies (Interagency Report, 2003), usefully summarises the keys to enhancing adaptive capacity:

- **Supporting sustainable livelihoods** by targeting development efforts at supporting communities' efforts to enhance social capital, preserve and restore natural capital, secure appropriate physical capital, enhance human capital and secure financial capital;

- **Ensuring equitable growth** by fostering growth in areas of the economy that provide opportunities for increased employment and higher returns for poor people's unique assets and contexts;
- **Improving governance** by making public institutions responsive, participative and accountable to those they serve in order to make decision-making processes and implementation activities robust and effective.

The specific elements of adaptive capacity is thus highly location-specific. Effectively enhancing adaptive capacity is heavily reliant on local knowledge and local involvement in design and implementation (Lim et al., 2005; Downing et al., 2005; Interagency Report, 2003). Indeed, "adaptation is largely a place-based activity, and a great deal of it can and should take place spontaneously or autonomously within those sectors and by those people, communities and enterprises most directly at risk" (Burton & van Aalst, 2004).

This highly location-specific nature of adaptation does not mean that adaptation can be wholly local, nor can it be undertaken without considerable external support. Adaptation is a combination of *implementation*, which is largely undertaken at the local level, and *facilitation*, which involves roles for actors from local to national to international levels. Facilitation includes creating an enabling environment in which the local implementation of adaptation is feasible. It also involves generating and sharing knowledge, building institutional capacity, carrying out coordination and management tasks and, not least, providing technological and financial resources.

The link to human development is central to this understanding of the term 'adaptation'. There is considerable overlap between efforts to enhance adaptive capacity and efforts to achieve broad human development goals. Like human development itself, effective adaptation will require not only improved understanding and more institutional capacity, but also significant financial resources. The World Bank estimates that the costs of 'climate-proofing' ODA (official development assistance), foreign direct investments and domestic investment in developing countries could range between $10 billion and $40 billion annually (World Bank, 2006a). Oxfam International (2007) estimates adaptation funding needs to be over $50 billion annually. The UNDP (2007) estimates $86 billion by 2015, and the UNFCCC Secretariat Dialogue Working Paper (UNFCCC Secretariat, 2007a; 2007b) offers an estimate of $49-171 billion annually by 2030. These estimates do not include a range of other adaptation costs related to building resilience and sustaining livelihoods in the face of climate change, which may not be direct capital investments, but which are critical elements of adaptation nonetheless.

In contrast to these estimates, the total amount of resources committed to adaptation (and other related preparatory activities such as the National Adaptation Programmes of Actions (NAPAs) through UNFCCC channels is dramatically smaller. It has come to less than $0.4 billion cumulatively over roughly the last ten years. A comparable amount is to come in the future from pending pledges plus the revenue expected through 2012 from the CDM adaptation surcharge.[1] Addressing the striking discrepancy in scale between the need and the supply of resources for adaptation is a key challenge of a future climate regime, and indeed was a focus of the Bali negotiations.

## 3. Current status of adaptation efforts

Initially, the UNFCCC was focused primarily on mitigation. Adaptation was considered less urgent as it was initially assumed that significant climate change impacts would not be evident for some decades. Today, the international negotiations appear to be looking at all the issues in the context of the four main strands identified for the dialogue on 'long-term cooperative action to address climate change' launched at COP 11 in December 2005: sustainable development, technology, adaptation and market opportunities. In reality, however, much of the negotiation effort and emphasis are still on mitigation, and burden-sharing is currently discussed only in the context of mitigation. Some analysts have informally observed that the adaptation regime is a decade behind the mitigation regime. They say that adaptation is at a comparatively embryonic stage

---

[1] The 2% charge on the CDM for the Adaptation Fund, and by COP 13 in December 2007 it had generated €37 million in resources, and is estimated by the World Bank to be able to generate $270-600 million (based on low and high 2012 estimates for CDM volume and prices). (See Steve Gorman, "Institutional Arrangements for Adaptation Fund: World Bank view", paper presented at the UNFCCC Workshop on the Adaptation Fund, 3 May 2006, Edmonton, Alberta, Canada.

characterised by institutional development, knowledge generation and preparation, which have to take place before on-the-ground progress can be made in communities suffering from climate impacts.

### 3.1 *Adaptation in the existing climate regime*

To the extent that adaptation efforts were initially envisioned in the United Nations Framework Convention on Climate Change (UNFCCC), they were embodied in the first meeting of the Conference of the Parties to the UNFCCC (COP 1) guidance to the Global Environment Facility (GEF) to initiate a three-stage process of short-, medium- and long-term support for adaptation. At present, the existing climate regime is channelling some resources towards adaptation through these initial GEF channels, as well as the three funds under the UNFCCC and the Kyoto Protocol: the Special Climate Change Fund (SCCF), the Least Developed Countries Fund (LDCF) and the Adaptation Fund.[2] Most of the resources result from a pledge made by the EU and other industrialised countries in 2001 when developing country support was needed to 'keep Kyoto alive'.[3] The resources committed through this pledge are channelled through the GEF, bilateral and multilateral ODA and the three funds. The fact that these resources are provided through diverse channels (including 'additional' ODA and the proceeds from the Clean Development Mechanism levy) makes it difficult to ascertain whether the pledging countries are meeting their commitments. Deriving financing from these diverse sources and for multiple purposes also makes it difficult to ascertain what proportion of the funds is in fact allocated to adaptation activities.

Increased attention to adaptation since COP 8 in New Delhi culminated in the adoption of a Five-Year Programme of Work on Impacts, Vulnerability and Adaptation to Climate Change at COP 11 in Montreal. This five-year work programme for the Subsidiary Body for Scientific and Technological Advice (SBSTA) is intended "... to improve understanding and assessment of impacts, vulnerability and adaptation, and to make

---

[2] Decision 7/CP.7 and decision 10/CP.7, respectively.

[3] The EU plus Canada, Iceland, New Zealand, Norway and Switzerland have pledged $410 million (€450 million at July 2001 exchange rates) per year by 2005 in climate change funding for developing countries.

informed decisions on practical adaptation actions and measures ..." [4] It left unspecified whether this five-year effort is focused exclusively on advancing *knowledge* (i.e. stocktaking, assessments, sharing of experiences, methodological development, data management and dissemination, etc.), or whether it allows for *implementation* of concrete adaptation measures.

The COP decision adopting the five-year programme specifies several 'implementation modalities' that do not entail the *implementation* of concrete adaptation measures, although the decision does leave room for the adoption of additional modalities that could include implementation. This issue was heavily debated at COP 11 and at the ensuing SBSTA24 in May 2006, which was given the task of further developing the modalities of the five-year programme, consistent with SBSTA's mandate to serve as an advisory body. The developing countries argued strongly in favour of an action-oriented approach that emphasises practical implementation and learning-by-doing. The course of the debates on the five-year programme closely mirrored the discussions surrounding the other sources of adaptation funding (GEF, SCC Fund, LDC Fund and Adaptation Fund) and resources for adaptation under the UNFCCC and Kyoto Protocol generally.

When the COP first introduced the Five Year Programme of Work (in Decision 1/CP.10), it also decided that the Global Environment Facility (GEF) should support implementation of pilot or demonstration projects (1/CP.10 7(b)(v) along with several other capacity-building and knowledge-enhancing activities. Until recently, the GEF had no explicit operational programme on adaptation, and its adaptation activities were secondary to its existing focal areas, which target global environmental benefits rather than locally needed adaptation. The three Marrakech funds are more directly targeted at adaptation activities, but the SCCF and the LDCF are funded through discretionary pledges, and the Adaptation Fund is subject to the uncertainty of the scale of the CDM and the market value of Certified Emissions Reductions (CERs).[5] A major decision at

---

[4] Decision 2/CP.11 taken at COP 11 (reported in FCCC/CP/2005/5/Add.1, 30 March 2006) (unfccc.int/resource/docs/2005/cop11/eng/05a01.pdf).

[5] It has been pointed out that the Adaptation Fund surcharge on the CDM is, ironically, a tax on developing country Parties insofar as it reduces both the volume of CERs traded (in favour of domestic reductions, emissions trading and

COP13/MOP 3 was, however, the agreement that the Adaptation Fund should be operationalized under a governing Board, with the Global Environment Facility (GEF) serving as its secretariat. This should enable the Adaptation Fund to move towards disbursing funds for adaptation activities. There was however no agreement on additional practical adaptation measures and the issue will be re-visited at SBSTA in June 2008.

The above points suggest that adaptation is an underdeveloped and fragmented part of the climate regime. The feeling, especially among developing countries, is that despite the widespread instances of climate impacts now being witnessed, the COP has not yet created sufficient elements of the climate regime aimed at implementing adaptation (Mace, 2005). One explanation offered for this situation is that the past decade has been a period of preparation. Implementation on the ground can only take place once this preparation has been done. Another reason is the small level of resources. Because contributions to adaptation funding to date have been voluntary and have not been directly tied to any particular underlying rationale, measurement of obligations or indicator of needs, they have thus been small in relation to the scale of the challenge. This is despite the fact that, by ratifying the UNFCCC, signatories have accepted a legally-binding commitment (within Article 4 of the UNFCCC, Article 11 of the Kyoto Protocol and in the Marrakech Accords) to provide financial resources in support of adaptation. Today, investment in adaptation *per se* is a minute fraction of the estimated need. The question remains as to whether adaptation funding based on voluntary pledges and a CDM levy can grow to the scale that is consistent with the adaptation challenge, and if not, then what type of framework would facilitate funding on the scale that is needed.

### 3.2 *Adaptation in mainstream development thinking*

Given that adaptation is intrinsically linked to the broader development agenda, efforts to address adaptation can only be successful if they are integrated into mainstream development thinking. While such integration[6]

---

Joint Implementation) and the CER revenue available for the countries hosting CDM.

[6] It is widely accepted that energy-based reductions can be effective only when mainstreamed into energy policy through comprehensive, sector-wide measures

has long been accepted in the case of mitigation, the same realisation has only dawned slowly when it comes to adaptation (Klein, 2006).

Although not much has happened in the recent past in terms of integrating adaptation into ODA (Agrawala et al., 2003 a-d; McGuigan et al., 2002; Burton & van Aalst, 2004; Klein et al., 2007; Klein, 2001), it is now emerging as something that policy-makers widely agree on. This was demonstrated at the 2006 meeting of the OECD Development Assistance Committee and the Environment Policy Committee, where the development and environment ministers of the OECD nations adopted a Declaration on Integrating Climate Change Adaptation into Development Cooperation.[7] The key element of the ministers' declaration was a commitment "to better integrate climate change adaptation in development planning and assistance". Simultaneously, the OECD development and environment ministers issued a Framework for Common Action Around Shared Goals (OECD, 2006b; 2006c), which reaffirms that threats to the environment have serious implications for poverty reduction and seeks to improve the coherence of efforts by development cooperation and environmental agencies in OECD countries in support of poverty reduction and the MDGs. Consistent with this approach, the Swedish Government announced at COP 13 in Bali the creation of an International Commission on Climate Change and Development, the purpose of which is to serve as a forum for sharing and operationalising of strategies on 'climate-proofing development'. Screening tools are now being developed in support of such work (e.g., Klein et al., 2007 and the OECD, 2007) has performed a stocktaking of mainstreaming efforts by donor agencies.

---

such as, for example, carbon taxes, renewable portfolio requirements and efficiency standards.

[7] Declaration on Integrating Climate Change Adaptation into Development Co-operation, adopted by Development and Environment Ministers of OECD Member Countries at the Meeting of the OECD Development Assistance Committee and the Environment Policy Committee, 4 April 2006, OECD, Paris (www.oecd.org/dataoecd/44/29/36426943.pdf). The ministerial declaration recognises that "responses to climate change should be coordinated with social and economic development in an integrated manner, taking into account the legitimate priority needs of developing countries for the achievement of sustainable economic growth and the eradication of poverty"(OECD, 2006a).

The World Bank (2006b) has elevated its attention to enhancing risk management approaches to enable development institutions and their partner countries to better address the growing risks from climate change and make current development investments more resilient to climate variability and extreme weather events. Many development agencies and NGOs (such as the Red Cross) are investing in efforts to learn from their long-standing experience with risk and crisis management of disasters, including those caused by extreme weather events, so that they can apply this knowledge to climate adaptation.

Ultimately, the purpose of integrating adaptation into development decision-making is to prevent climate change from being separated from the other stresses and trends that affect development as experienced by vulnerable citizens. However, it is important to note that development is primarily driven by the vast majority of investments that are *not* part of ODA. Non-ODA foreign and domestic investment in developing countries (~\$1.6 trillion/yr) is roughly two orders of magnitude larger than ODA. Integrating adaptation into mainstream development, therefore, will ultimately involve not just influencing ODA investment decisions, but those much larger non-ODA investments as well. This ambitious objective implies not just a change in thinking, but changes in terms of institutional capacity, policy coherence and human and financial resources.

## 4. An adaptation regime: Looking forward

Given the magnitude of the climate change challenge ahead, significant resources – probably in the range of tens to hundreds of billions of dollars per year – will be needed in the future to cover costs related to building adaptive capacity and implementing adaptive responses in developing countries. This will inevitably entail transfers from industrialised to developing countries. A key question is how to obtain the level of political acceptability needed to create and expand the financial flows required to support developing country adaptation efforts (Parry et al., 2005). Various explanations can be put forward as to why the EU and other industrialised parties would choose to provide substantial resources for implementing adaptation in developing countries.

First, there is a political argument for funding adaptation as a means of engaging developing countries. It is thought that extending meaningful adaptation assistance to developing countries will help to encourage them to participate in mitigation aspects of the climate regime as well.

Developing countries have unambiguously staked out a position that they are entitled to funding for adaptation and that it should be provided by industrialised countries, and that without this funding a more substantive form of participation in the mitigation aspects of the regime would be unlikely. Indeed, little further progress in international negotiations can be expected without taking these concerns into account (Ott et al., 2004).

The second explanation, like the first, is also based on enlightened self-interest. This argument justifies industrialised countries funding of adaptation as a means of ensuring that developing countries remain viable partners for economic growth, global governance and international security. Climate change without proactive adaptation could arguably cause significant damage to the economies and governance systems of developing countries, and perhaps ultimately generate a flow of environmental refugees (Byravan & Rajan, 2005; 2006; Barnett & Adger, 2006; Sachs, 2007).

The third explanation has its origins in the ethical and legal context in which the debate about climate change is unfolding. It argues that industrialised countries have a responsibility to contribute resources towards adaptation in developing countries. It is discussed in more detail in the following section, as a) it carries the greatest sense of inevitability as a motivator of EU investment in developing country adaptation, b) it may offer some quantitative measure of the obligation industrialised parties may have to fund adaptation in developing countries, and c) it helps in the elaboration of fundamental principles consistent with the Climate Convention around which a future adaptation regime can be envisioned.

## 4.1 *Fundamental characteristics of adaptation*

This section sets out three fundamental characteristics of adaptation that provide a framework within which proposals for a future adaptation regime can be assessed.

### 4.1.1 *The link to human development*

The first fundamental characteristic derives from the intrinsic connection between effective adaptation and human development. As discussed in section 8.2, adaptation must include both adaptive responses and enhancement of adaptive capacity. The most effective adaptation strategy in the long-term is to provide secure livelihoods, foster equitable economic growth and improve governance. As articulated in the Interagency Report

(2003), "adaptation requires the development of human capital, strengthening of institutional systems, and sound management of public finances and natural resources. Such processes build the resilience of countries, communities and households to all shocks and stresses, including climate variability and change, and are good development practice in themselves."

If an adaptation regime is to induce the most effective adaptation, then a major component of its objective must be identifying and promoting those activities that can most effectively enable and accelerate these "good development practices".

### *4.1.2 Responsibility for climate change and climate impacts*

The second characteristic relates to the scale of the adaptation challenge, not just in terms of creating institutions and generating knowledge, but in terms of additional costs that will have to be borne, either to make adaptation possible or to compensate damages. (See above estimates of additional costs of adaptation.) The need for such resources inevitably raises the question of the *responsibility* of nations for causing climate change and, in turn, damages caused by this climate change. As Baer (2006) writes:

> That it is wrong to harm others (or risk harming them) for one's own gain, and that one owes compensation if one does such harm, is as close to a universal ethical principle as exists. It is a principle that is justified in all kinds of ethical or moral frameworks, from divine revelation to deontological ethics to social contract theory and others. It is in fact a prime example of what some philosophers call common (or commonsense) morality.

Over time this common moral principle has become firmly encoded in national case law and legal reasoning with respect to environmental pollution within national boundaries. International law, too, echoes this same principle. The Stockholm Declaration of 1972 declares in the famous Principle 21 (reaffirmed in Principle 2 of the Rio Declaration) that states have "the responsibility to ensure that activities within their jurisdiction or control do not cause damage to the environment of other States or of areas beyond the limits of national jurisdiction" and reiterates in Principle 22 that "States shall cooperate to develop further the international law regarding liability and compensation for the victims of pollution and other environmental damage caused by activities within the jurisdiction or control of such States to areas beyond their jurisdiction" (UN Conference on the Human Environment, 1972).

The UNFCCC, however, specifically avoids establishing legal responsibility for climate damage,[8] although it does potentially lay the basis for recognising state-based responsibility in a manner that is reiterated in the Kyoto Protocol and the Marrakech Accords. UNFCCC Annex II countries are obligated to assist developing countries in meeting adaptation costs under specified circumstances (as detailed in Articles 4.3, 4.4, 4.8, 4.9 and 4.11). In particular, Article 4.3 commits UNFCCC Annex II countries to "provide new and additional resources to meet the agreed full incremental costs of implementing measures…" including "preparing for the adaptation to the impacts of climate change". In addition, Article 4.4 states that UNFCCC Annex II countries "shall also assist the developing country Parties that are particularly vulnerable to the adverse effects of climate change in meeting costs of adaptation to those adverse effects". Some have interpreted this latter clause as "an implicit acceptance of responsibility for causing climate change" but this has not been unambiguously established.[9] The Kyoto Protocol (Article 11) further requires that the "implementation of these existing commitments shall take into account the need for adequacy and predictability in the flow of funds and the importance of appropriate burden sharing among developed country Parties".

---

[8] As Tol & Verheyen (2004) discuss, the industrialised nations successfully resisted any codification of state-based responsibility for compensation for climate damages at the time of the UNFCCC negotiations. In response, and in anticipation of the future evolution of the climate regime or legal action, several member states of the Alliance of Small Island States (AOSIS) issued a declaration that "…signature of the Convention shall in no way constitute a renunciation of any rights under international law concerning state responsibility for the adverse effects of climate change…" Such a clause was in fact unsuccessfully proposed for inclusion in the UNFCCC text itself (see submissions of Vanuatu on behalf of AOSIS, Elements for a Framework Convention on climate change, in a set of informal papers provided by delegations, related to the preparation of a framework convention on climate change, UN Doc. A/AC.237/Misc.1/Add3).

[9] Tol & Verheyen (2004) quoting P. Sands (1992), "The United Nations Framework Convention on Climate Change", *RECIEL* (1:3).

### *4.1.3 Facilitating and implementing local solutions*

The third fundamental characteristic is the fact that effective adaptation, like human development itself, is ultimately a highly localised process that defies generic solutions. Vulnerable communities and households suffer from constellations of stresses with complex and location-specific interactions, and the interventions that build resilience tend to be correspondingly location-specific. As discussed in section 2, effective adaptation therefore requires both *facilitation* and *implementation* of location-specific efforts to adapt to environmental change.

While implementation of adaptation is largely undertaken at the local level, facilitation relies on actors at local to national to international levels. Generating and sharing knowledge across these scales is challenging. Even more challenging is to mobilise resources and efficaciously distribute them to local levels. But successful strategies have evolved over time within the development community. These tend to be participatory, stakeholder-driven undertakings. They are often based on the close involvement of local civil society organisations, place strong value on local information and knowledge, and invest heavily in building local trust in and acceptance of the undertakings as well as building durable links between the different levels (Lim et al., 2005, for example). As explained in the Interagency Report (2003), "In Southern Africa numerous adaptation techniques are used by poorer farmers to deal with anticipated drought. These include water and soil management techniques, resistant crop varieties and food production methods. However, these techniques are often known only locally, or to certain ethnic groups."

The challenge then for an adaptation regime is to move between the local, regional and national levels and effectively channel knowledge and resources to the local sites where they are needed for the implementation of adaptation.

## *4.2 A climate regime to support effective adaptation*

It remains unclear how adaptation will be addressed within the UNFCCC. What is clear, though, is that a regime that successfully manages to support adaptation will need to overcome the three challenges implied above: i) to respect the close link between adaptation and development, ii) to create an acceptable framework for recognising responsibility for generating adaptation funds, and iii) to develop instruments that support the objectives of facilitating and implementing local adaptation activities via good coordination between the local, regional and national levels.

Importantly, in order for legally binding commitments to be accepted, an adaptation regime will need to inspire a high level of confidence that adaptation will be supported in a manner that is *effective* and *efficient.* Securing adaptation funding will not be easy, and, without significant assurances that resources would be efficiently and equitably allocated to adaptation activities, those funds will not be forthcoming. The notions of effectiveness and efficiency cover several elements:

- Efficient institutions and mechanisms, which allow resources to contribute directly to adaptation without their being wasted on bureaucracy or diverted due to corruption.
- Assurance that resources are being put towards *bona fide* adaptation activities rather than activities that would be taking place anyway. (That is not to say that scaling up of existing activities could not be legitimate adaptation.)
- Governance structures must enable equitable decision-making and instil trust in the governments that are providing resources for adaptation, the governments receiving resources and the communities that need the resources.
- Developing countries in need of adaptation will need to build sufficient absorptive capacity to make effective use of adaptation funding. In many countries, this level of absorptive capacity does not exist today and will not exist without improvements in knowledge and expertise, institutional strength, good governance, transparency and accountability.
- A specific legal instrument for adaptation must not 'ghettoise' adaptation, and make it harder, rather than easier, to ultimately integrate adaptation into mainstream development activities (as well as other multilateral environmental agreements and trade agreements).

With these points in mind, we now consider each of the three fundamental characteristics of adaptation in turn with respect to their ramifications for an adaptation regime.

i) **The link to human development.** In today's regime, UNFCCC Annex II countries have sought some assurance that the money they provide is going towards new activities in developing countries rather than 'business as usual' activities directed towards existing development objectives.[10] One can anticipate that this concern with 'additionality' will only grow more pressing if the climate regime progresses beyond the current stage in which commitments to finance adaptation efforts are voluntary and contributions are donated under the banner of philanthropy, and evolves into a regime in which Parties accept legally-binding commitments to contribute resources towards adaptation. However, given that a major objective of an adaptation regime is to build adaptive capacity, which entails investments in fundamental improvements in livelihoods, equitable growth and democratic governance, then it might prove counterproductive if UNFCCC Annex II countries cling to the additionality construct with respect to adaptation activities. Firstly, as argued in section 8.2, it will be a burdensome and unrewarding challenge to attempt to distinguish many adaptation activities *per se* from broad-based development activities in general. Secondly, attempts to demonstrate additionality would be likely to channel attention towards those adaptation activities that may be the most easily discernible as additional, but not necessarily the most important urgent, or cost-effective contributors to adaptive capacity. This phenomenon - sometimes referred to as the 'paradox of additionality' (Sugiyama & Michaelowa, 2001) – favours separate and immediate project-based activities, often focused on well-defined marginal improvements to localised infrastructure rather than broad-based and long-term activities that may much more effectively address the underlying drivers of vulnerability and build adaptive capacity. As discussed by Klein et al. (2007), programme budget support may ultimately be more efficient and effective than project-based investments.

---

[10] The most explicit example of this is the CDM in which mitigation must be 'additional' to ongoing technological and sectoral change. In the adaptation realm, it manifests itself as the requirement for 'incremental cost' accounting.

ii) **Responsibility.** In order to generate sufficient funding for adaptation or, in the words of the Convention, to ensure 'adequacy and predictability' and 'appropriate burden sharing' in funding adaptation, the climate discussion will have to evolve from general acknowledgement of ethical and legal principles to specific definitions of responsibility and their quantification. One example of fairly concrete methods is the typology of quantitative indicators of responsibility recently produced by the World Resources Institute (Baumert & Markoff, 2003). A fairly large and developed body of work exists exploring options for explicitly linking commitments to specific quantifiable indicators of responsibility (AOSIS, 1991; Brazil, 1997; Ott et al., 2004; Government of Tuvalu, 2005; Oxfam International, 2007; Baer et al., 2007)[11] As adaptation funding requirements are assessed, with due account taken of the magnitude of anticipated climate change impacts, the obligation to provide funding would then be allocated to Parties according to an agreed definition of responsibility.[12]

There should be no illusions that a discussion of responsibility will be easy. Industrialised Parties will continue to be extraordinarily wary of any obligations that commit them to providing significant additional adaptation funds. Moreover, it would be unwise to establish any system to generate more financial resources for adaptation than could be absorbed. Yet initiating the discussion will still be useful. It would encourage industrialised country Parties to articulate very clearly the conditions under which they could be confident that adaptation resources would be used effectively. More generally, it would advance the global dialogue, which began at the global climate change summits in Stockholm and Rio de Janiero, on

---

[11] Negotiating an acceptable indicator of responsibility will be challenging. Parties can be expected initially to put forward proposals that favour their interests, and rationalise them as "appropriate burden sharing" in keeping with "their common but differentiated responsibilities". In addition, Parties may need to resolve issues such as the attribution of emissions to importers or exporters and the distinction between 'luxury' and 'subsistence' emissions.

[12] One can also anticipate this leading, as in the case of mitigation, to issues concerning compliance.

responsibility for trans-boundary environmental impacts. Moreover, early and effective deployment of resources for adaptation in developing countries will lessen the likelihood that, in the longer term, significant damages will lead to legal claims for compensation.

iii) **Local solutions with accountability.** Any attempt to use adaptation funds to facilitate and implement dispersed and tailored activity at the local level will bring with it the challenge of addressing the need for transparent management and accountability. It will be necessary to ensure accountability to those providing the funding as well as to the recipient communities. Industrialised countries will not willingly fulfil their responsibilities to provide resources for adaptation if the recipient countries do not also transparently uphold their responsibility to efficiently and effectively undertake the intended adaptation. An adaptation regime involving legally-binding commitments would therefore need to allow for oversight and ensure accountability, while avoiding micro-management and keeping decision-making authority and discretion within the host countries and communities.

The balance of control between the Party providing the funds and the Party (or community) receiving the funds will be a challenging one to strike. From the perspective of the Party providing the funds, relinquishing any control may be unappealing and may undermine their willingness to accept an adaptation regime. On the other hand, from the perspective of the Party receiving the funds, these funds may be considered obligatory payments as a result of ongoing climate change damage caused by the Parties that emit high levels of pollution, and thus control should rightfully be in the hands of the recipient. The basic tension is between an implicit charity mentality versus a perspective of compensation or liability. To date, a mutually acceptable balance has arguably still not been struck (Mace, 2005; Mönher & Klein, 2007; Müller, 2006a; 2006b).

### *Institutional mechanisms*

Various institutional mechanisms are being considered for administering and delivering adaptation activities. Four major ones are:

- to expand ODA infrastructure to accommodate adaptation
- to create or extend a globally centralised adaptation fund

- to create locally-focused funds ("Autonomous Adaptation Funds"[13])
- to establish an insurance mechanism for adaptation[14]

These options will undoubtedly be fleshed out in more detail as more is learned from on-the-ground experiences about effective methods for delivering adaptation. Each of them will have its strengths and weaknesses with respect to the three fundamental principles outlined above for effective adaptation. Some of these can be anticipated.

With respect to the link to development, the first and third of these four options provides a built-in link by definition, whereas the connection is somewhat tenuous for the second and, especially, the third options.

With respect to responsibility, it would in theory be possible to explicitly link obligations for funding to countries' responsibility, regardless of whether the funding is then channelled as additional appropriations for a country's ODA budget, replenishment schedules for a globally centralized fund or local funds, or premiums for insurance mechanisms. The question, then, is how to do this in a rigorous, transparent, and principled way.

With respect to local solutions, while all the options can be tailored to local needs in theory, the third option is only one that is structured around the notion of locally appropriate delivery. The ODA system has become increasingly capable, over the years, of internalising 'bottom-up', locally-appropriate responses, but it is still maturing in this area and still has its serious critics.

---

[13] For a discussion, see for example: African Association for Public Administration and Management AAPAM, Nairobi, and the Dag Hammarskjöld Foundation, Uppsala, Sweden under the title "Autonomous Development Funds in Africa. Report of the Expert Consultation on the role of Autonomous Funds as intermediaries in channelling money for social and economic development in Africa." Kampala, Uganda, 4-6 April 1995. A similar construct, which was put in place in various Latin American states as a way of minimising the negative impacts of structural adjustment programmes in the 1990s (and continues to operate today), has also been identified as a viable mechanism for achieving adaptation (Cruz, 2005).

[14] See, for example, AOSIS (1991), Mills & Lecomte (2006), Germanwatch (2005), Bals et al. (2005) and Linnerooth-Bayer & Mechler (2006).

While much more can be said and much more needs to be learned, consistency with the three underlying principles of adaptation should serve as a criteria in assessing the appropriateness and viability of proposed institutional mechanisms for delivering adaptation.

## 5. Closing comments

In summary, a climate regime that attempts to effectively address adaptation is faced by key set of issues that derive from the very nature of adaptation.

- *How can the link to human development be internalised into adaptation efforts?* There are intrinsic similarities between activities that enhance adaptive capacity and those that advance basic development. This suggests that it may be counterproductive to attempt to draw a firm distinction between adaptation projects and other development efforts. It also suggests that there are significant lessons to be transferred from the long history of ODA, such that investments in building adaptive capacity can start immediately.
- *How and to whom should responsibility be attributed?* A certain amount of climate change is now inevitable, and adaptation to limit the damages will require considerable resources, especially in developing countries. Industrialised countries, with their GHG emissions per capita much in excess of the global average, have a certain legal and moral responsibility to provide resources, as well as pragmatic reasons for wanting to do so.
- *How can adaptation needs be identified and prioritised in a manner that respects the local nature of adaptation?* Implementing adaptive responses is a very location-specific endeavour, implying that the intimate involvement of local communities is a key ingredient for effective implementation. At the same time, activities at higher levels (sub-national, national, or international) can provide a facilitate role, in order to enable successful local adaptation.
- *What would be effective and appropriate institutions or mechanisms for achieving adequate levels of adaptation?* In other words, what institutions or mechanisms would effectively take into account the intrinsic links with development, the twin objectives of facilitation and implementation of local responses, and the need to provide adequate resources by relating funding commitments to national responsibility.

The scientific and empirical findings that have led to an increased sense of urgency with regard to mitigation are now also precipitating a more urgent response on adaptation. This chapter has focussed on issues related to constructing an adaptation regime with the aim of supporting and achieving adaptation in developing countries at a level that is commensurate with the challenge. We have sought to set out some of the relevant background, including three fundamental characteristics of adaptation for consideration when attempting to define a viable regime.

## References

Agrawala, S., A. Moehner, A. Hemp, M.K. van Aalst, S. Hitz, J. Smith, H. Meena, S.M. Mwakifwamba, T. Hyera and O.U. Mwaipopo (2003a), *Development and Climate Change in Tanzania: Focus on Mount Kilimanjaro*, Environment Directorate and Development Cooperation Directorate, Organisation for Economic Cooperation and Development (OECD), Paris.

Agrawala, S., T. Ota, A. U. Ahmed, J. Smith and M.K. van Aalst (2003b), *Development and Climate Change in Bangladesh: Focus on Coastal Flooding and the Sunderbans*, Environment Directorate and Development Cooperation Directorate, Organisation for Economic Cooperation and Development (OECD), Paris.

Agrawala, S., T. Ota, J. Risbey, M. Hagenstad, J. Smith, M.K. van Aalst, K. Koshy and B. Prasad (2003c), *Development and Climate Change in Fiji: Focus on Coastal Mangroves*, Environment Directorate and Development Cooperation Directorate, Organisation for Economic Cooperation and Development (OECD), Paris.

Agrawala, S., V. Raksakulthai, M.K. van Aalst, P. Larsen, J. Smith and J. Reynolds (2003d), *Development and Climate Change in Nepal: Focus on Water Resources and Hydropower*, Environment Directorate and Development Cooperation Directorate, Organisation for Economic Cooperation and Development (OECD), Paris.

Agrawala, S. (ed.) (2005), *Bridge Over Troubled Waters: Linking Climate Change and Development*, Environment Directorate and Development Cooperation Directorate, Organisation for Economic Cooperation and Development (OECD), Paris.

Alvarenga, K. and N. Eckley Selin (2005), "Report of the UNFCCC Seminar on the Development and Transfer of Environmentally Sound Technologies for Adaptation to Climate Change, 14-16 June 2005", *Earth Negotiations Bulletin*, Vol. 12, No. 271, International Institute for Sustainable Development, Canada.

AOSIS (Alliance of Small Island States) (1991), also published as Chapter 7 of Peter Hayes and Kirk Smith (eds), *The Global Greenhouse Regime: Who Pays?*, London: Earthscan, 1993.

Baer, P. (2006), "Adaptation to Climate Change: Who Pays Whom?", in N. Adger, S. Huq, J. Paavola and M.J. Mace (eds), *Justice in Adaptation to Climate Change*, Cambridge, MA: MIT Press.

Baer, P., T. Athanasiou and S. Kartha (2007), "The Right to Development in a Climate-Constrained World: The Greenhouse Development Rights Framework", Berlin: Heinrich Böll Stiftung (available at www.ecoequity.org/GDRs).

Bals, Christoph, Ian Burton, Sonja Butzengeiger, Andrew Dlugolecki, Eugene Gurenko, Erik Hoekstra, Peter Höppe, Ritu Kumar, Joanne Linnerooth-Bayer, Reinhard Mechler and Koko Warner (2005), "Insurance-Related Options For Adaptation To Climate Change", Executive Summary, The Munich Climate Insurance Initiative (MCII).

Barnett, Jon and W. Neil Adger (2006), "Climate Change, Human Security, and Violent Conflict", *Global Environmental Change and Human Security Newsletter*, No. 1, Spring.

Baumert, Kevin A. and Matthew Markoff (2003), Indicator Framework Paper, World Resources Institute, Washington, D.C. (available at http://cait.wri.org).

Brazil (1997), The Brazilian Proposal, Paper 1 in "Implementation of the Berlin Mandate: Additional proposals from Parties Addendum; Note by the secretariat", Document FCCC/AGBM/1997/MISC.1/Add.3 (available at unfccc.int/cop3/resource/docs/1997/agbm/misc01a3.htm).

Burton, Ian and Maarten van Aalst (2004), "Look Before You Leap: A Risk Management Approach for Incorporating Climate Change Adaptation in World Bank Operations", World Bank, Washington, D.C.

Byravan, Sujatha and Sudhir Rajan (2005), "Immigration could ease climate-change impact", *Nature*.

Byravan, Sujatha and Sudhir Rajan (2006), "Providing New Homes for Climate Change Exiles", *Climate Policy* (6).

Canada (2005), "Comments Received from Canada, September 16, 2005", in Comments received from participants on the revised programming document for GEF-4, (GEF/R.4/9) (GEF/R.4/Inf.12, 23 September 2005), as quoted in Mace (2005).

Cruz, Oscar Nelson (2005), "Social Investment Funds to Help Communities Reduce Disaster Risk And Facilitate Climate Adaptation", presentation at Workshop on Financial Services and Community Adaptation, 2nd Red Cross/Red Crescent International Conference on Climate Change and Disaster Risk Reduction, 22-23 June, The Hague.

Downing, T.E. and A. Patwardhan with R.J.T. Klein, E. Mukhala, L. Stephen, M. Winograd and G. Ziervogel (2005), "Assessing Vulnerability for Climate Adaptation", in B. Lim, E. Spanger-Siegfried, I. Burton, E. Malone and S. Huq (eds), *Adaptation Policy Frameworks for Climate Change: Developing Strategies, Policies and Measures*, Cambridge: Cambridge University Press.

Germanwatch (2005), "Insuring the Uninsurables. Preliminary thoughts regarding the basic structure of an insurance related mechanism (IRM)", Draft report, Germanwatch, Bonn.

Gorman, Steve (2006), "Institutional Arrangements for Adaptation Fund: World Bank view", presentation given at the UNFCCC Workshop on the Adaptation Fund, 3 May, Edmonton, Alberta, Canada.

Government of Tuvalu (2005), "Seminar of Government Experts: Draft Future Actions Strategy - Tuvalu", presentation made at the Seminar of Government Experts, 16-17 May, Bonn, Germany.

Interagency Report (2003), *Poverty and Climate Change: Reducing the Vulnerability of the Poor through Adaptation*, Inter-agency report by the African Development Bank (AfDB), Asian Development Bank (ADB), Department for International Development (DFID, UK), Federal Ministry for Economic Cooperation and Development (BMZ, Germany), Ministry of Foreign Affairs - Development Cooperation (DGIS, The Netherlands), Organisation for Economic Cooperation and Development (OECD), United Nations Development Programme (UNDP), United Nations Environment Programme (UNEP) and the World Bank.

IPCC (2007), Climate Change Impacts, Adaptation and Vulnerability: The Contribution of Working Group II to Climate Change 2007, Fourth Assessment Report of the Intergovernmental Panel on Climate Change.

Klein, Richard J.T. (2001), Adaptation to Climate Change in German Official Development Assistance – An Inventory of Activities and Opportunities, with a Special Focus on Africa, Deutsche Gesellschaft fur Technische Zusammenarbeit: Eschborn, Germany.

Klein, Richard J.T. (2006), "Mainstreaming Adaptation to Climate Change into Official Development Assistance: Promoting Synergies or Diverting Money?" *Global Environmental Change and Human Security Newsletter*, No. 1, Spring.

Klein, Richard J.T., S. Eriksen, L. Naes, A. Hammill, C. Robledo and K. O'Brien (2007), "Portfolio Screening to Support the Mainstreaming of Adaptation to Climate Change into Development Assistance", *Climatic Change* (forthcoming).

Lim, Bo, Erika Spanger-Siegfried, Ian Burton, Eizabeth Malone and Saleemul Huq (2005), *Adaptation Policy Frameworks for Climate Change Developing Strategies, Policies and Measures*, Cambridge: Cambridge University Press.

Linnerooth-Bayer, Joanne and Reinhard Mechler (2006), "Insurance for Assisting Adaptation to Climate Change in Developing Countries", *Climate Policy* 6 (6), pp. 621-636.

McGuigan, C., R. Reynolds and D. Wiedmer (2002), Poverty and Climate Change: Assessing Impacts in Developing Countries and the Initiatives of the International Community, London School of Economic Consultancy Project for the Overseas Development Institute.

Mace, M.J. (2005), Funding for Adaptation to Climate Change: UNFCCC and GEF Developments since COP-7, RECIEL 14 (3).

Mills, E. and Eugene Lecomte (2006), "From Risk to Opportunity: How Insurers Can Proactively and Profitably Manage Climate Change" (http://www.ceres.org/pub/publication.php?pid=0).

Möhner, A. and R.J.T. Klein (2007), *The Global Environment Facility: Funding for Adaptation or Adapting to Funds?*, Climate and Energy Working Paper, Stockholm Environment Institute, June.

Müller, B. (2006a), "Climate of Distrust: The 2006 Bonn Climate Change Adaptation Fund Negotiations" (available at http://www.oxordclimatepolicy.org).

Müller, B. (2006b), "Adaptation Funding and the World Bank Investment Framework Initiative", background report prepared for the Gleneagles Dialogue Government Working Groups, Mexico, 7-9 June.

OECD (2006a), *Declaration on Integrating Climate Change Adaptation into Development Co-operation*, adopted by Development and Environment Ministers of OECD Member Countries at the Meeting of the OECD Development Assistance Committee and the Environment Policy Committee at Ministerial Level, 4 April, OECD, Paris (www.oecd.org/dataoecd/44/29/36426943.pdf).

OECD (2006b), *Framework for Common Action Around Shared Goals*, adopted by Development and Environment Ministers of OECD Member Countries at the Meeting of the OECD Development Assistance Committee and the Environment Policy Committee at Ministerial Level, 4 April, OECD, Paris (http://www.oecd.org/dataoecd/44/27/36427017.pdf).

OECD (2006c), *Putting Climate Change Adaptation in the Development Mainstream*, Policy Brief, OECD, Paris, March.

Ott, H.E., H. Winkler, B. Brouns, S. Kartha, M.J. Mace, S. Huq, Y. Kameyama, A.P. Sari, J. Pan, Y. Sokona, P. Bhandari, A. Kassenberg, E. Lèbre La Rovere and A.A. Rahman (2004), *South-North Dialogue on Equity in the Greenhouse: A proposal for an adequate and equitable global climate agreement*, study commissioned by the German Federal Ministry for Economic Cooperation and Development (available at http://www.wupperinst.org/download/1085_proposal.pdf).

Oxfam International (2007), *Adapting to Climate Change: What's Needed in Poor Countries, and Who Should Pay*, Oxfam Briefing Paper 104, Oxfam International Secretariat, Oxford.

Parry, J., A. Hammill and J. Drexhage (2005), *Climate Change and Adaptation: Summary*, International Institute for Sustainable Development, Canada.

Patt, A.G., R.J.T. Klein and A.C. de la Vega-Leinert (2005), "Taking the uncertainty in climate-change vulnerability assessment seriously", *Comptes Rendus Geoscience*, 337(4).

Smit, B. and O. Pilifosova (2001), "Adaptation to climate change in the context of sustainable development and equity", in J.J. McCarthy, O.F. Canzianni, N.A. Leary, D.J. Dokken and K.S. White (eds), *Climate Change 2001: Impacts, Adaptation, and Vulnerability*, Contribution of Working Group II to the Third Assessment Report of the Intergovernmental Panel on Climate Change, Cambridge: Cambridge University Press.

Smith, J.B., R.J.T. Klein and S. Huq (2003), *Climate Change, Adaptive Capacity and Development*, London: Imperial College Press.

Sugiyama, T. and A. Michaelowa (2001), "Reconciling the Design of CDM with Inborn Paradox of Additionality Concept", *Climate Policy*, Vol. 1, No. 1, January.

Tol, R.S.J. and R. Verheyen (2004), "State Responsibility and Compensation for Climate Change Damages - A Legal and Economic Assessment", *Energy Policy*, 32 (9).

Turner, B.L.I., R.E. Kasperson, P.A. Matson, J.J. McCarthy, R.W. Corell, L. Christensen, N. Eckley, J.X. Kasperson, A. Luers, M.L. Martello, C. Polsky, A. Pulsipher and A. Schiller (2003), "A framework for vulnerability analysis in sustainability science", *Proceedings of the National Academy of Sciences* 100, 8074-8079.

UNDP (2007), *Human Development Report 2007/2008 – Fighting Climate Change: Human Solidarity in a Divided World*, Basingstoke: Palgrave Macmillan.

UNFCCC (2006), "Background paper on Overview of Possible Institutional Options for the Management of the Adaptation Fund", prepared for the UNFCCC Workshop on the Adaptation Fund, Edmonton, Canada, 3-5 May.

UNFCCC Secretariat (2007a), "Report on the analysis of existing and potential investment and financial flows relevant to the development of an effective and appropriate international response to climate change", Dialogue Working Paper No. 8, submitted to the Dialogue on long-term cooperative action to address climate change by enhancing implementation of the Convention Fourth workshop, Vienna, 27–31 August.

UNFCCC Secretariat (2007b), "Background paper on analysis of existing and planned investment and financial flows relevant to the development of effective and appropriate international response to climate change" (available at unfccc.int/files/cooperation_and_support/financial_mechanism/application/pdf/background_paper.pdf).

World Bank (2006a), *Clean Energy And Development: Towards An Investment Framework*, World Bank , 5 April.

World Bank (2006b), *Managing Climate Risk - Integrating Adaptation into the World Bank Group Operations*, World Bank Group Global Environment Facility Programme, report written by Maarten van Aalst.

# 8. The Dual Markets Approach to Reducing Emissions from Deforestation and Degradation in Developing Countries

*Ned Helme, Diana Movius, Matt Ogonowski & Jake Schmidt*[*]

## 1. Introduction

Tropical deforestation accounts for approximately 20% of global carbon dioxide emissions.[1] Modelling by the European Commission estimates that limiting global temperature increases to less than 2°C with a 50% probability through cost-effective measures requires net global emissions from land use changes to fall to near zero by 2020. In Montréal in 2005, the United Nations Framework Convention on Climate Change (UNFCCC) began a series of workshops to discuss potential post-2012 policy instruments to reduce emissions from deforestation in developing countries. UNFCCC workshops on the subject were held in Rome in 2006

---

* Ned Helme, President and founding member of the Center for Clean Air Policy (CCAP), Washington D.C. Diana Movius is a Policy Associate, Matt Ogonowski a Senior Policy Analyst and Jake Schmidt, International Program Manager, all at the Center for Clean Air Policy (CCAP). The authors would like to acknowledge Pedro Barata and Mac Wubben (Center for Clean Air Policy), Thelma Krug (Secretary of Climate Change and Environmental Quality, Brazil), Artur Runge-Metzger (European Commission) and participants in the Deforestation Emissions Reduction work group of the Dialogue on Future International Actions to Address Global Climate Change.

[1] World Bank (2007).

and in Cairns, Australia in March 2007. The climate negotiations at COP-13 in December 2007 led to the adoption of a pilot work programme for further methodological work to reduce emissions from deforestation and forest degradation. This programme will focus on assessments of changes in forest cover and associated green house gas emissions, methods to demonstrate reductions of emissions from deforestation and the estimation of the amount of emission reductions from deforestation. The issue will continue to play a large role in the negotiations stemming from the Bali Roadmap.

In this chapter, the Center for Clean Air Policy (CCAP) presents the 'dual markets approach' for reducing emissions from deforestation and degradation (REDD). This approach creates a new market for emissions reductions from deforestation and degradation, not directly linked with the post-2012 global carbon market. Annex I countries would meet a portion of their post-2012 greenhouse gas reduction target through this REDD-only market. The COP would set a maximum on the percentage any single Annex I country can achieve through overseas REDD. Annex I countries would specify at the outset which developing countries' deforestation and reduction programmes they will invest in to achieve their REDD target. A pre-2012 preparation phase for developing countries – to include emissions inventories, capacity building and pilot projects – will allow developing countries to design effective programmes for the post-2012 REDD market.

This new proposal is presented in the context of REDD policy discussions underway in the UNFCCC. Various structures have been proposed for a mechanism for providing positive incentives to reduce emissions from deforestation and degradation, and over 60 countries and 15 non-governmental organisations have submitted position papers to the UNFCCC. The chapter first reviews important issues any REDD mechanism must address and lays out the specific proposals endorsed by Partie and then moves on to two general approaches to incentives for REDD that have shaped the discussions: the integrated market approach (where REDD reductions would be integrated directly into the post-2012 global carbon market) and the fund-based approach (dependent on official development assistance or institutional financing). The major advantage of the market approach is its ability to mobilise large amounts of funding; its downside is the potential to disrupt the global carbon market that is already functioning well. A fund approach would avoid disruption to the carbon market, but would likely fail to mobilise sufficient financial resources to reduce the deforestation significantly.

The Dual Markets Approach offers many of the advantages of these previous proposals while at the same time avoiding their major drawbacks. A separate REDD market assists learning-by-doing, allowing the REDD market to mature and stabilise before linking with the post-2012 global carbon market. The Dual Markets Approach would still leverage private-sector resources, while minimising the risk of destabilising the post-2012 global carbon market. Specific REDD targets adopted by Annex I countries would provide a minimum demand for developing countries' REDD credits and would therefore facilitate planning of their national REDD programmes. Setting a maximum ceiling on Annex I REDD purchases would safeguard global stabilisation goals in case REDD reductions prove unreliable.

While creating a new dual market system inevitably creates some risk, these are no larger than the risks of integrated-market or fund-based approaches. CCAP puts forth the dual markets approach as a workable solution for the post-2012 commitment period that can bridge the gap between competing proposals and enable effective action on REDD.

## 2. Cross-cutting issues

To be effective, any new programme to reduce deforestation and degradation must address the following key issues:

*Financial incentives for participation by developing and developed countries.* Appropriate incentives are needed to encourage developed countries to contribute financial resources. Annex I countries would either purchase credits in a market system or provide funds in a non-market system. Developing countries would need appropriate rewards for changed land-use.

*Reference rate.*[2] The development of national reference rates for deforestation is the foundation of any international approach to REDD. Reference rates will require inventories of historical forest cover and carbon sequestration since 1990. Brazil has proposed that future reductions in deforestation and emissions be estimated through comparison with observed historical levels. Others have proposed using 'virtual' baselines developed as counterfactual business-as-usual trends in future years. To prevent domestic leakage, baselines would ideally be national.

---

[2] This can also be referred to as a baseline, base year or base period level.

*Funding implementation and capacity-building in developing countries.* This is necessary pre-2012 in order to provide a foundation for effective post-2012 action on REDD.

*Mitigation costs and reduction potential.* Countries will require detailed knowledge of the costs and reduction potentials of mitigation options before investing in REDD actions. Accurate information will help establish credibility of the system in the eyes of developed countries.[3]

*Leakage.* If not designed appropriately, a REDD mechanism could simply shift deforestation from one area to another. Unchanged international demand for certain commodities, such as palm oil or beef, could lead to deforestation in new areas not under government REDD policies.

*Net versus gross deforestation.* The mechanism could support reductions in either net or gross deforestation. Gross refers to emissions from deforestation only, whereas net includes sequestration (for example, soils and regrowth). Rewards based on gross deforestation reductions help to ensure that replacement of cleared primary forests by cash crop plantations or reforestation projects are not credited. Net accounting would guard against inadvertently crediting countries where emissions have actually increased due to forest fires, logging, etc.

Most proposals for REDD seek to address these issues in some way. The options for generating financing have proven to be among the most debated elements. We review the various financing proposals below, including specific submissions of parties, advantages and disadvantages.

---

[3] Research in this area remains preliminary, but over the past few years some estimates of the costs of avoiding deforestation have been developed. For example, one study predicts that in 2025, measures costing <\$6, <\$14 and <\$27 per tonne $CO_2e$ could sequester a total of 44, 98 and 137 Gt $CO_2e$, respectively (Sohngen, 2005). Another study projects that at \$27 per tonne, tropical deforestation could nearly be stopped by 2055. This would require payments of \$465-660 per hectare – about \$2.5 trillion total (Sohngen & Beach, 2006). In addition, an analysis conducted for the *Stern Review of the Economics of Climate Change* concluded that the opportunity costs of avoided deforestation in eight countries (Bolivia, Brazil, Cameroon, Congo, Ghana, Indonesia, Malaysia and Papua New Guinea) would range from \$3 billion to \$6.5 billion annually, with \$5 billion being perhaps the most realistic estimate (Grieg-Gran, 2006).

## 3. Financing REDD: Review of approaches

The majority of REDD proposals submitted to the UNFCCC fall under two general categories: market and fund-based. A market-based approach would allow developing countries undertaking voluntary actions that reduce their deforestation rate or maintain carbon stocks to generate carbon credits, which they can sell at a market-determined price per tonne $CO_2e$ reduced. This market mechanism would be linked to a 'post-2012' carbon market as an extension of the existing Kyoto carbon market. Credits generated from REDD actions would be equivalent to 'post-2012' carbon credits (e.g. those generated through the CDM) and could be traded along with or in place of such credits.[4]

Non-market approaches would rely on contributions to a fund or funds from developed country governments and sources such as through official development assistance (ODA), international financial institutions and the private sector. For the REDD mechanism, financing is then distributed from this fund to activities that either reduce deforestation or reward countries for successful forest protection. For example, Brazil proposed a strict fund-based approach that would provide direct payments to countries for carbon reductions achieved through REDD measures, but not generate carbon credits. Other types of funds have also been proposed, such as to cover REDD capacity-building activities and project implementation.[5]

Parties have proposed the following financing mechanisms and funds:

*The REDD market mechanism* would operate in the existing Kyoto Protocol carbon market. Crucial components include credits from reduced deforestation emissions tradable with developed countries and CDM credits, deeper targets by Annex I and additional crediting for early action (e.g. pre-2012 pilot projects, capacity-building, and monitoring). Reductions

---

[4] For example, Compensated Reduction has been proposed as a market-based mechanism of rewarding actions that reduce deforestation rates by allowing the trading of reductions in the carbon market (Santilli et al., 2005).

[5] To date, most governments are in favour of some combination of market-based support and funds. One example is the proposal of the Coalition for Rainforest Nations which combines a market-based reward system for reducing deforestation with separate funds for capacity-building and pilot projects.

would be measured against a reference scenario determined by a reference emissions rate and a development adjustment factor. Market mechanisms are proposed by the Coalition for Rainforest Nations[6] and supported by most developing countries.

A non-market *REDD fund*, proposed by Brazil, would reward countries that demonstrate reductions in their deforestation rate through a direct funding mechanism. The funds would be distributed in proportion to the emission reductions achieved, measured against a reference scenario comprised of an historic national baseline. Funds would be awarded only after a reduction was demonstrated from this baseline, as opposed to a projected baseline.

A non-market *stabilisation fund*[7] would finance efforts to maintain, or stabilise, currently forested areas. This applies to countries with historically low rates of deforestation, to prevent the potential deforestation of their existing forests. This fund could be generated through taxes or levies on the carbon market, or through voluntary contributions.

A non-market *enabling fund*[8] would assist REDD infrastructure development activities in non-Annex I countries (e.g. capacity-building, monitoring programme development, establishment of inventories and baselines). This fund would provide early financial resources to countries in need of infrastructure development, so that these countries could eventually participate effectively in a REDD mechanism.

## 4. Advantages and disadvantages of market and fund-based approaches

An important advantage of the carbon market approach is the capability of providing large sums of money to support reducing deforestation[9] by

---

[6] This includes Bolivia, Central African Republic, Costa Rica, Democratic Republic of Congo, Dominican Republic, Fiji, Ghana, Guatemala, Honduras, Kenya, Madagascar, Nicaragua, Panama, Papua New Guinea, Samoa, Solomon Islands and Vanuatu (joint submission February 2007 available at www.unfccc.int).

[7] Also proposed by Coalition for Rainforest Nations.

[8] Also proposed by Coalition for Rainforest Nations.

[9] For example, one presentation at the Cairns Workshop (Noble, 2007) outlined that $2-25 billion per year may be required for REDD incentives, while current non-market based financing is significantly lower than this amount.

creating a direct channel for private sector investments. A market-based system would also be self-functioning while carbon caps are maintained and would empower many public and private sector players. A market system would likely generate lower-cost REDD credits that can be purchased as offsets by developed countries.

However, the global carbon market could be flooded with a large volume of low-cost REDD credits, creating volatility or drops in carbon market prices. A strict carbon market approach will naturally incentivise the lowest price emissions reductions; in the long term, addressing the more difficult (and higher opportunity cost) drivers of deforestation may require higher cost incentives than the early stages of the REDD-included carbon market mechanism would generate.[10]

The main advantage of the non-market (fund-based) approach relates to protecting the CDM. Because REDD credits would not enter the carbon market, the non-market REDD mechanism would better maintain the integrity of the post-2012 global carbon market. A fund proposal would make REDD actions more dependent on government leadership by reducing private sector forces. The largest drawback is that funding would depend upon voluntary contributions from Annex I countries or institutions, and would therefore not likely produce levels of funding comparable to market approaches. This approach lacks a crediting component that would incentivise sufficient Annex I contributions, making the REDD fund less appealing than the carbon market.

## 5. The dual markets approach: CCAP's new proposal for REDD

In this section we suggest an alternative proposal for supporting REDD actions – the dual markets approach. This proposal would create a new market for the sale of REDD credits post-2012, separate from the existing carbon market. The COP would establish a maximum percentage of the overall national carbon target that countries could choose to achieve through the new REDD market. Annex I countries could then elect to meet

---

[10] This is likely only a concern to the extent that not providing incentives for the higher opportunity cost deforestation actions would 'lock-in' a certain level of deforestation trends. For example, investments in agricultural activities that lead to deforestation could be more difficult (and costly) to reverse once the infrastructure, political support, etc. have been generated.

up to that percent of their post-2012 reduction target through purchasing REDD credits from the new REDD market. The proposal also highlights a pre-2012 preparatory phase to better enable developing countries to participate in the post-2012 REDD market.

*Pre-2012 capacity building and inventory development.* Both developed and developing countries would lay the groundwork for an effective REDD mechanism. Annex I countries would commit to providing ODA and other financing to fund key activities including the establishment of developing country Land Use, Land-Use Change and Forestry (LULUCF) emissions inventories, capacity-building, research studies of potential mitigation measures and costs, and implementation of pilot projects.[11]

*Post-2012 new REDD market.* The cornerstone of the dual markets proposal is the creation of a new REDD-only carbon market. This market is separate from the traditional (Kyoto/CDM) carbon market. The REDD-only market would begin in 2012 with the next commitment period. In this new REDD market, reductions generated through REDD actions in developing countries would be sold to Annex I countries to satisfy those countries' REDD commitments. Emission reductions from REDD activities in developing countries could not be substituted for traditional carbon reductions (e.g. CDM, emissions trading or emissions reducing actions) beyond the level established as that Annex I country's REDD commitments.

*Overall target-setting.* The COP would first decide the maximum percentage of GHG reductions that Annex I countries can achieve through the REDD market for the first post-2012 commitment period. This can be viewed as a ceiling on purchases from the REDD market. The maximum is necessary to: 1) ensure that steady progress toward global stabilisation goals is not undermined by over-reliance on less-certain REDD credits, 2) allow the REDD market time to develop before integrating with the already mature Kyoto market and 3) avoid the risk of REDD credits flooding the post-2012 global carbon market.

*Annex I target setting.* Individual Annex I countries would then commit to meeting a certain percentage of their own reduction target through $CO_2$ equivalent reductions from REDD activities in developing countries. This percentage would be up to the maximum established by the

---

[11] A pre-2012 phase has been noted as important in the majority of Party submissions.

COP. The Annex I REDD target would be a firm commitment, serving essentially as both a minimum and a maximum.[12] Take as an example the current EU proposal for a 30% reduction in emissions by 2020.[13] Under the REDD dual markets approach, the EU would allocate a share of this target (for example, 5%) to be met through the REDD market. The remaining 25% would then be achieved through domestic GHG mitigation or purchases in the traditional post-2012 carbon market.[14]

*Signalling strong demand for REDD activities.* At the beginning of the post-2012 phase, developing countries would draw upon their experience in the preparatory pre-2012 phase to estimate the likely quantity and cost of reductions. They would draft land management plans to achieve these reductions. Annex I buyers would review these proposals and determine from which developing countries' programmes to eventually buy. Developing countries therefore would know at the outset of a commitment phase which Annex I countries will buy their credits. This gives developing countries a minimum level of demand for REDD credits, and generates incentive to act early and develop effective programmes in order to recruit Annex I buyers. These agreements could take the form of bilateral contracts, including for example a certain amount of credits at a certain price per tonne, with built-in pricing mechanisms such as escalators linked to the global carbon price.

*Ensuring emissions reductions goals through non-REDD carbon credits or borrowing.* If the ex-post reductions from REDD actions are lower than anticipated at the end of the first commitment phase, Annex 1 countries could move a portion of their REDD targets to traditional Kyoto caps. In other words, they could meet the remainder of their REDD targets through carbon market purchases or use any banked carbon allowances. Annex I countries would also have the option to borrow against their future REDD commitments. This flexibility provides options for Annex I countries if

---

[12] Annex I countries could meet their REDD targets through use of government funds, mandates on companies (e.g. through targets in domestic emissions trading systems) or a combination.

[13] Contingent on participation by other developed countries.

[14] During the introductory post-2012 phase of the REDD mechanism, developed countries would not be allowed to increase their REDD share, in order to ensure the integrity of each market.

REDD shortfalls occur, while still ensuring that the overall global emissions reduction goals are eventually met.

*Review by the COP.* The dual markets approach specifies that Annex I REDD targets stay constant for the upcoming post-2012 commitment period, and the COP determines the maximum percentage allowed from REDD. Constant percentages, with an emphasis on longer commitment phases (e.g. 2012-20 or 2012-30) could encourage real, long-term action. However, the COP would periodically review the dual market system, taking into account the quality and dependability of REDD credits and the stability of the REDD market. The COP can make needed changes to the system, such as altering percentages allowed from REDD, potentially permitting Annex I increases in their share and eventually determining whether to link with the global carbon market.

## 6. Questions and answers

The CCAP Dual Markets Approach is a realistic strategy that circumvents many disadvantages of previous proposals. This question and answer section aims to clarify the logic of this new approach for those new to this approach and to the sinks policy field. It also addresses the practical functioning of the dual markets approach.

*Why is a separate REDD market necessary? Why is it necessary for the COP to set a maximum (ceiling) on REDD credits?* For the next commitment phase (2012-20 or 2012-30), a separate REDD market with a maximum cap has many advantages. These stem from the fact that most developing country REDD programmes and policies would be new and untested at such a large scale. It may be difficult to determine precisely how many carbon tonnes can be sequestered from developing countries' new REDD policies until those policies have had time to demonstrate results. A maximum set on REDD credits therefore provides some safety net for global stabilisation goals if REDD programmes produce 'hot air', or fail to produce reductions. The maximum prevents Annex I countries from relying too much on reductions from untested REDD programmes for the first commitment phase.[15] The maximum therefore emphasises the

[15] The COP can establish a review board, like the CDM board, to add flexibility if needed.

importance of domestic reductions by developed countries and safeguards demand for already-tested reductions such as the CDM.

A separate REDD market avoids major disruption of the post-2012 global carbon market. In a pure integrated-market approach such as compensated reduction, fluctuations in REDD supply and pricing would directly affect CDM prices, which could cause overall volatility in the carbon market that is currently functioning well. A dual markets approach keeps separate an emerging market (REDD) from the more mature carbon market until questions have been resolved.[16] This approach allows the REDD market, and individual countries' REDD programs, time to develop before integration with the existing global carbon market.[17] Developing countries can craft and test functional REDD policies in a time frame appropriate to national circumstances. With countries testing national REDD programs for the first time, it is unlikely they would produce a large surplus of reductions that would exceed the maximum placed by the COP.[18]

*In a dual system, how will REDD market affect the existing carbon market?* In the dual markets approach, REDD credits would probably affect carbon prices the most at the end of the commitment phase. If REDD tonnes fail to materialise sufficiently by the end of the first commitment phase and many Annex I countries use the carbon market to offset REDD shortfalls, carbon prices would rise at the end of the post-2012 commitment phase. This proposal mitigates the effect of REDD tonnes on the existing carbon market, compared to entirely integrated market approaches. Nonetheless, the option to achieve part of the national target through REDD will impact carbon prices to some degree throughout the commitment period.

*How will the private sector participate?* This approach will facilitate the participation of the private sector. Annex I commitments for REDD will not

---

[16] As discussed earlier, linking of the two markets could be decided eventually by the review system of the COP.

[17] Successful precedents for creating dual environmental markets already exist – in the United States, for example, national and state-level fuel economy standards and renewable portfolio standards (RPS), i.e. renewable quotas have functioned effectively along with independent emissions trading programmes for GHGs and criteria pollutants.

[18] At a minimum for the upcoming commitment phase.

be official development aid; rather, countries can structure their commitment in a variety of market-friendly ways. First, governments could translate a REDD commitment into mandates on companies, such as company targets in an emissions trading scheme. A government could create an allowance system to meet their post-2012 targets through a cap and trade system (e.g. the EU emissions trading scheme). Private companies could be required to invest in the REDD market to achieve a specified level of reduced deforestation, or could be given this option up to a ceiling percentage. In this case, the sum of all companies' REDD contributions would add up to a portion of the total REDD target of the country. Trading between the companies could also be allowed or encouraged. Under this scheme companies could meet a portion of their carbon responsibility through REDD investments and would also help that country meet their post-2012 REDD target.

A creative way a country could leverage private sector finance to help meet a REDD target is included in the United States Senate's Lieberman-Warner Climate Security Act of 2008, which would set aside 2.5% of allowances in a US cap and trade system for REDD and other forestry projects.[19] Proceeds from the sale of 2.5% of the allowances would ostensibly create a fund that he United States would use to invest in developing country REDD programmes. These would not serve as direct offsets for the industrial sector and would instead create additional reductions above and beyond the domestic reductions achieved under cap and trade. With proper carbon accounting it would be possible to measure the amount of reductions afforded by this sum of money. Annex I countries could estimate ahead of time the revenue needed, and the associated portion of allowances needed, for REDD projects in order to meet that Annex I country's post-2012 REDD goal. This approach would rely on the private sector as well, as their payments for auctioned allowances would comprise the financial source for the REDD funding.

*Would Annex I countries be allowed to increase their REDD targets during the commitment period?* Annex I countries would not be allowed to increase their REDD targets during the first commitment period, but can adjust their

---

[19] The Lieberman-Warner Climate Security Act of 2008 (S. 2191), as reported by the Environment and Public Works Committee on 5 December 2007, is the leading climate change bill working its way through the United States Congress.

REDD targets for the second commitment period. To address uncertainties in a new REDD market, the dual markets approach allows borrowing into the next commitment period. If these markets were not kept separate, countries could do more in the REDD market as a way to take greater advantage of borrowing and thus erode the level of reductions achieved in the next commitment phase.

*Why is it necessary for Annex I countries to decide at the outset which developing countries' credits they will buy?* Slowing deforestation requires considerable change and long-term planning to create such programmes, often at a significant cost. Developing countries will need some guarantee that there will be a buyer for their REDD credits. To determine REDD programme structure and implementation, host countries will also need to have some estimate of funds they will eventually receive.

*How can the pre-2012 phase be financed?* Inventories, capacity-building and implementation will be financed by multilateral institutions and official development aid from Annex I governments.[20] However, once countries and investors know that a REDD market will exist, some incentives should emerge for contributions to capacity-building. The ability to meet a portion of national GHG reduction targets through cheaper overseas REDD credits should encourage Annex I countries; reductions will also incorporate on-the-ground projects, which are ripe for private-sector investment. One creative approach could be policy-based lending, which is a technique used by international finance institutions. Such loans (at a very low interest rate) cover the salaries or fees of experts in the host country to develop and implement policies.

## 7. Conclusion

The development of a framework to encourage actions to reduce GHG emissions from deforestation and degradation in developing countries is crucial in order to address global climate change. With approximately 20% of carbon dioxide emissions globally coming from tropical deforestation, REDD actions on a broad scale will be necessary to stabilise GHG concentrations in the atmosphere. A REDD mechanism will need to

---

[20] The World Bank, for example, has started the Forest Carbon Partnership Facility, and the government of Australia has committed funding for forest conservation in the Asia-Pacific region.

address many issues, including establishing reference scenarios and baselines, creating effective incentives for developing and developed countries, and ensuring a sufficient and consistent flow of funding. Parties have submitted a variety of proposals for an international REDD mechanism, relying on carbon market integration and/or voluntary contributions.

The dual markets approach would combine critical elements of previous market and fund-based approaches and address many shortcomings. By creating a separate REDD market, in contrast to other market-based proposals, the dual markets framework would maintain the integrity of the existing carbon market, while generating more financing than a fund-based approach. Annex I countries would agree to specific REDD targets, providing a guaranteed level of demand for developing country REDD credits while safeguarding stabilisation goals against potential uncertainties of REDD reductions. The proposal would provide a range of options to both developing countries and Annex I countries, including voluntary participation, borrowing from future REDD targets, making up REDD shortfalls with carbon market credits, and review by the COP to consider linking to the post-2012 carbon market. The dual markets approach would facilitate a system of learning-by-doing for the upcoming post-2012 commitment period and allow the international community to craft a workable incentive system to reduce emissions from deforestation.

## References

Grieg-Gran, M. (2006), "The Cost of Avoiding Deforestation", report prepared for the *Stern Review of the Economics of Climate Change*, International Institute for Environment and Development, London, October.

Houdashelt, M., J. Schmidt, M. Ogonowski and D. Movius (2007), "Positive Trade Incentives for Encouraging Greater Participation in a post-2012 Climate Agreement", Working Paper prepared for the CCAP Dialogue on Future International Actions to Address Climate Change, Center for Clean Air Policy, Washington, D.C., June.

The Lieberman-Warner Climate Security Act of 2008 (S. 2191), 110th Congress 2008 (see www.Thomas.gov, the information website of the US Library of Congress).

Noble, I. (2007), "Financing Reduced Emissions from Deforestation and Degradation", presentation at the UNFCCC Workshop on Reducing Emissions from Deforestation in Developing Countries, Cairns, Australia, 7-8 March (http://unfccc.int/files/methods_and_science/lulucf/application/pdf/070307noble.pdf).

Santilli, M. et al. (2005), "Tropical deforestation and the Kyoto protocol", in Paolo Moutinho and Stephan Schwartzman (eds), *Tropical Deforestation and Climate Change*, Instituto de Pesquisa Ambiental da Amazônia and Environmental Defense.

Schmidt, J., N. Helme, J. Lee and M. Houdashelt (2006), *A Sector-based Approach to the Post-2012 Climate Change Policy Architecture*, Future Actions Dialogue Working Paper, Center for Clean Air Policy (CCAP), Washington, D.C. (http://www.ccap.org/international/future.htm).

Sohngen, B. (2005), "Marginal Cost Curves for Carbon Sequestration in Forests: Estimates for Boreal, Temperate, and Tropical Regions of the World" (http://aede.osu.edu/people/sohngen.1/forests/ccforest.htm).

Sohngen, B. and R. Beach (2006), "Avoided Deforestation As a Greenhouse Gas Mitigation Tool: Economic Issues for Consideration" (http://aede.osu.edu/people/sohngen.1/forests/ccforest.htm).

United Nations Framework Convention on Climate Change (UNFCCC) (2007), submissions from Parties: Views on the range of topics and other relevant information relating to reducing emissions from deforestation in developing countries (http://unfccc.int/documentation/documents/advanced_search/items/3594.php?rec=j&priref=600004203#beg).

World Bank (2007), "Forest Carbon Partnership Facility", Concept Note, Washington, D.C., March (http://carbonfinance.org/Router.cfm?Page=FCPF&FID=34267&ItemID=34267&ft=DocLib&ht=47&dtype=25617&dl=0).

# 9. THE CRITICAL ROLE OF TECHNOLOGY FOR INTERNATIONAL CLIMATE CHANGE POLICY

*CAROLYN FISCHER, CHRISTIAN EGENHOFER & MONICA ALESSI**

## 1. Introduction

While there are different opinions about whether a stabilisation of greenhouse gas (GHG) emissions in line with the United Nations Framework Convention on Climate Change's (UNFCCC) objective can be reached with technically proven technology,[1] no one questions the need to develop new and technically unproven (i.e. breakthrough) technologies in the long term, and certainly beyond 2050. Unless a magic-bullet solution for reducing emissions can be found, a portfolio of technologies will be needed. Among these, wind, solar and biofuels are growing rapidly, albeit from a small base. Other technologies, such as hydrogen, are considered promising, but face substantial challenges in terms of cost and large-scale implementation. As fossil fuels are expected to dominate the world's energy supply portfolio for some decades, carbon capture and storage (CCS) may become an important bridge before we enter into a low-carbon future. Over and above inevitable technical hurdles, the scale of the task means that widespread global deployment of technologies, however

* Carolyn Fischer is Research Fellow in the Energy and Natural Resources Division of Resources for the Future, Washington, D.C.; Christian Egenhofer is Senior Research Fellow at CEPS and Monica Alessi is Research Fellow at CEPS.

[1] Pacala & Socolow (2004) argue that the climate problem for the next 50 years could be solved with current technologies, whereas Hoffert et al. (2002) hold that new and revolutionary technologies would be needed.

promising, will take decades before the cumulative effect of investments makes a substantive contribution to combating climate change.

Such massive technological change depends on the right combination of public R&D investments ('technology push') and policies to provide economic incentives for private-sector innovation and widespread technology deployment ('market pull'). A global climate change agreement can influence the full spectrum of the process of technological change. Commitments to GHG abatement policies are the source of market-pull signals that encourage innovation and deployment of low and near-zero carbon technologies. Commitments to R&D efforts represent an option for enhancing and coordinating technology-push policies, basically on three levels: support for basic research, mostly done by academic institutions; component testing in test facilities and laboratories; and full-scale demonstration, an important pre-requisite for market push. Given the primacy of technological advance in tackling the climate problem, a principal question for negotiators is how a global agreement affects technology development and diffusion.

## 2. The economics of R&D and innovation

In the case of climate change, twin market failures are at play, inhibiting the emergence of sufficient incentive to produce technological solutions on their own.[2] The most immediately relevant market failure is the fact that the costs of global warming are not borne directly by GHG emitters, which leads to fossil-fuel prices that are 'too cheap' and a level of GHG emissions that is too high from a societal perspective. The economist's policy prescription is to 'put a price' on GHG emissions – for example through a GHG tax or cap-and-trade system – thereby forcing individuals and firms to internalise the cost that they are placing on everyone else when they emit GHGs. In addition, there also are market deficiencies related to the development and adoption of new technologies. These technology market problems are not as relevant for environmental problems addressed over the course of years as they are for climate policy developing over decades or centuries and requiring much more dramatic changes in technology.

---

[2] This section draws on the summary in de Coninck et al. (2007).

Jaffe et al. (2005) identify the following relevant types of technology market imperfections.[3]

- First, due to 'knowledge spillovers', innovating firms cannot keep other firms from benefiting from their new knowledge and, therefore, cannot capture for themselves all the benefits of innovation. In addition, the process of competition typically will drive a firm to sell a new device at a price that captures only a portion of its full value. While patents and other institutions are employed to protect firms' investments in innovation, such protection is inherently imperfect.[4]
- Second, 'adoption spillovers' may be relevant in the adoption and diffusion of new technology. For a number of reasons, the cost or value of a new technology to one user may depend on how many other users have adopted the technology. In general, users will be better off the more other people use the same technology, so there is a benefit associated with the overall scale of technology adoption ('network externalities'). The supply-side counterpart, learning-by-doing, describes how production costs tend to fall as manufacturers gain production experience. If this learning spills over to benefit other manufacturers without compensation, it can represent an additional adoption externality. Finally, network externalities exist if a product becomes technologically more valuable to an individual user as other users adopt a compatible product (as with telephone and computer networks). These phenomena can be critical to understanding the existing technological system, forecasting how that system might evolve, and predicting the potential effect of some policy or event.
- Third, market shortcomings arise through incomplete information. While all investment is characterised by uncertainty, the uncertainty associated with the returns to investment in innovation often is particularly large. Potential returns also are asymmetrically distributed and the developer of new technology typically is in a better position to assess its potential than others and may find investors sceptical about promised returns. In the context of

---

[3] See Jaffe et al. (2003) for an overview of issues at the interface of environmental policy and technological change.

[4] An opposing incentive of conferring monopoly rights to an innovator may induce overinvestment in redundant research efforts, as firms race to get the patent.

environmental problems such as climate change, the huge uncertainties surrounding the future effects of climate change and the magnitude of the policy response and, thus, the likely returns to R&D investment, exacerbates this problem further. Another type of information problem relates to the inability of current policy-makers to credibly commit to a long-term emissions path. As a result, the long-term price signal associated with GHG reductions is likely to be significantly diminished relative to what it would need to be in order to achieve significant future reductions.

- Finally, incomplete information lies at the source of principal-agent problems, as when a builder or landlord chooses the level of investment in energy efficiency in a building, but the energy bills are paid by a later purchaser or a tenant.

The fact that markets underinvest in new technology (Jaffe et al., 2005) strengthens the case for making sure that environmental policy is designed to foster, rather than inhibit innovation. In cases where environmental costs have not been fully priced, it also is likely that the rate of investment in such technology is significantly below the socially desirable level. And it is unlikely that environmental policy alone can create sufficient incentives.

It is a basic principle of economics that sound policy-making requires at least as many types of policy instruments as there are market problems to be addressed (Tinbergen, 1956). Hence, the optimal set of climate policies likely includes instruments explicitly designed to foster innovation, and possibly technology diffusion, in addition to GHG emissions policies that stimulate new technology as a side effect of internalising the GHG externality. Likewise, long-term technology R&D alone is not sufficient because it provides no direct incentives for adoption of new technologies and because it focuses on the longer term, missing near-term opportunities for cost-effective emissions reductions (Philibert, 2003; Sandén & Azar, 2005; Fischer & Newell, 2004).[5]

---

[5] In addition to setting the proper stage for private investment, governments also have their own investments to manage. Public infrastructure is particularly important, as it has a long life span and predetermines people's choices of where to live and work, what to consume, what sort of economic activities to carry out and of other people to communicate with. Infrastructure development will be a critical

## 3. Technology and developing countries

Arguably, the producers of new technologies are most likely to be the industrialised countries, although this is also true for some fast-growing developing countries (Brewer, 2007). Long-term success requires that effective R&D and emissions policies are put in place. In developed countries, that policy strategy will most effectively be a mix of largely market-based measures to overcome the market failures and smart public investments. However, the consumers of climate-friendly technologies will be all countries, including developing and transition economies. In these countries, poverty, limited institutional capacity, governance problems and other issues loom larger as barriers than the market failures discussed earlier. As a result, it is naïve to expect the market-based strategies for emissions reductions and innovation to generate the same results in these circumstances. Information exchange systems are particularly valuable in this context: not only do they contribute to accelerated technology deployment, but they also increase awareness of climate change and low-carbon technology solutions. They may also enhance endogenous capacities and technologies in developing countries. For LDCs, access to technologies is a key issue, while for some fast-growing developing countries concessional finance is needed for wider technology cooperation.

## 4. Investment and finance

In a strategy to stabilise $CO_2$ at about 450 ppm in the low-carbon scenario, the World Bank has identified an investment need for non-OECD countries of around $165 billion p.a. for electricity generation, with current private and public sector resources funding around half.[6] To fill that gap, three sources of funding for mitigation are available: voluntary actions; international grants, e.g. GEF; and carbon trading. Carbon trading is likely to confer the biggest flow of funds to developing countries - between $20 and $120 billion per year - but it requires a long-term global regulatory framework (i.e. a 2050 target) with differentiated responsibilities and

---

factor for both costs and overall effectiveness of climate policies. Moreover, history matters. Past infrastructure investment determines the present, although economists disagree on how widespread this path-dependency is.

[6] World Bank reports to the Development Committee in the spring and fall of 2006 on the Clean Energy Investment Framework - now developed into an action plan.

intermediate targets. New financial instruments are required, especially to ensure market continuity in the post-2012 period.

International financial institutions can be an important source of finance and policy and technical advice regarding the financing of investments needed for clean energy for development. Also the creation of new financing instruments e.g. for clean energy, power rehabilitation, project development or venture capital could support this effort in several possible ways (World Bank, 2006 and 2007).

## 5. Technology approaches within the UNFCCC and beyond

Under Article 4(5) of the UNFCCC, industrialised countries have a special obligation to promote, facilitate and finance the transfer of technology to developing countries. The focus in the UNFCCC has thus primarily been the transfer of technology to developing countries for energy efficiency, and lately for adaptation. Various processes have been introduced. The Expert Group on Technology Transfer (EGTT) is the main instrument. It addresses technology needs assessments, information exchange, enabling activities and capacity-building. Two finance mechanisms that encourage technological diffusion through financial assistance and cooperation should be mentioned: the Global Environmental Facility (GEF) and the Clean Development Mechanism (CDM). In addition to technology transfer being addressed within the UNFCCC framework, there have been 'pull approaches' to technological change within industrialised countries such as the Climate Technology Initiative (CTI), the Gleneagles Plan of Action (GPOA), the Asia-Pacific Partnership on Clean Development and Climate (APP), the Major Economies Meeting (MEM) or the idea of global sectoral industry approaches.

### *5.1 The Expert Group on Technology Transfer*[7]

The Expert Group on Technology Transfer (EGTT) was established by the Marrakech Accords, with the objective of enhancing the implementation of Article 4(5) of the Convention, by analysing and identifying ways to facilitate and advance technology transfer activities and making

[7] The authors would like to acknowledge useful comments and information provided by Bernard Mazijn, Member of the UNFCCC EGTT, Vice-Chair in 2005, Chair in 2006 for this section.

recommendations to the Subsidiary Body for Scientific and Technological Advice.

In collaboration with other agencies such as the UN Industrial Development Organisation (UNIDO), UN Environment Programme (UNEP), UN Development Programme (UNDP), Global Environmental Facility (GEF) and the Climate Technology Initiative (CTI), the Expert Group on Technology Transfer has worked for six years on developing the key elements for the development and transfer of technology, namely technology information (via a clearinghouse and networking with regional technology information centres), technology needs assessments, enabling environments and mechanisms.

During the last years, two new areas of work have been added: Technologies for adaptation and innovative options for financing.

Technology and technology transfer appeared in the majority of decisions arising from COP-13 in Bali. A high-level workshop was held on technology cooperation during the conference. A strategic programme was launched to encourage concrete demonstration projects, to create more attractive environments for investment and to provide incentives to the private sector for technology transfer. This will be set up through GEF with international financial institutions and representatives of the private financial sector.

After much debate, the mandate of the EGTT was extended for an additional five years. Particular attention will be paid to the assessment of gaps and barriers to the use of and access to financing resources as well as the development of performance indicators that can be used to regularly monitor and evaluate progress on the development, deployment and transfer of environmentally sound technologies.

### 5.2 *The Global Environmental Facility*

The UNFCCC's central financial tool for technology transfer is the GEF. It is the main provider of grants for environmental projects in developing countries, of which climate change claims 40% of the current yearly budget. The Marrakech Accords in 2001 clarified the role of the fund and created three specific new funds to generate the conditions and leverage for private financing, including technology transfer for adaptation and mitigation purposes: The Trust Fund, the Special Climate Change Fund (SCCF) and the Least Developed Countries Fund (LDCF). The management of the adaptation fund has been agreed upon at COP-13 in Bali.

### 5.3 *The Clean Development Mechanism*

Although the CDM does not have an explicit technology transfer mandate, it partly contributes to technology transfer by financing a number of emissions reduction projects that use technologies not yet available in the host country. Based on a survey of participants in CDM projects, Haites et al. (2006) find that 33% of project developers claim that technology was actually transferred, referring, according to this definition, to the transfer of equipment for the most part, but also with some involvement of knowledge transfer. CDM project participants also claim that technology transfer represents 66% of the estimated annual emissions reductions. Interestingly, projects claiming technology transfer are often substantially larger projects, often involving more than one country. Four countries - Brazil, China, India and South Korea - account for 67% of the projects claiming technology transfer and for 75% of the annual emissions reductions. Haites et al. also point out that the governments of China, India and South Korea mention technology transfer as a key eligibility criterion for CDM projects, implying that the CDM can be an important vehicle for technology transfer if the terms of project eligibility (i.e. explicitly including technology transfer) and administrative efficiency are guaranteed.

The CDM could in particular become a vehicle for sector-based projects for technology transfer in developing countries. There are good arguments in favour of an extension towards a broader range of forestry and bio-energy projects and the development of the CDM into a more programmatic or sectoral crediting mechanism although many technical issues would need to be solved first. More importantly, massive expansion of the CDM or other credits needs to be matched by *demand*, most of which is currently provided by the EU ETS or other Kyoto Protocol signatories.

There are a number of design features of the CDM that limit its scope and which have been highlighted, among others, in an ECP report (Egenhofer et al, 2005) on CDM. "As a market-based instrument, the CDM tends to favour projects that are cheapest and most secure. This has put countries with weak economies and a lack of adequate institutional infrastructure in a disadvantageous position to benefit from the CDM". This weakness is confirmed by Haites et al. (2006): "This limitation is a barrier to avoid fossil-fuel-based development in countries in earlier stages of economic development." To reduce the imbalances, compensatory actions may be needed, such as the use of different financing instruments (e.g. ODA and financing from international finance institutions). Another limitation of the CDM is the focus on limited types of gases, which excludes opportunities for a range of technologies to be developed and transferred.

### 5.4 *Beyond the UNFCCC*

While the UNFCCC's focus has been on the transfer of technologies to developing countries, the industrialised member countries of the UNFCCC initiated a process focused on a 'pull approach' to technological change. The oldest of such initiatives emerged within the UNFCCC itself in 1995: The Climate Technology Initiative (CTI), which is implemented through the International Energy Agency (IEA). The IEA was later tasked by the Gleneagles G8+5 process[8] with analysis, planning and knowledge dissemination during the implementation of the Gleneagles Plan of Action. Within the Gleneagles process, the focus has enlarged from technology transfer to all stages of the deployment of new technologies, from RD&D to commercial deployment.

Other initiatives include the Asia-Pacific Partnership on Clean Development and Climate Change (APP), the Major Economies Meeting (MEM) and global sectoral industry approaches.

- The *Asia-Pacific Partnership on Clean Development and Climate Change (APP)* was created in 2006 by Australia, China, India, Japan, South Korea and the US, and is now joined by Canada. It is a public-private partnership on a regional scale, encouraging interaction of business, government and researchers from partner countries and designed to facilitate the development, diffusion, deployment and transfer of cost-effective, cleaner and more efficient technologies and practices among the partners through concrete and substantial cooperation aimed at achieving practical results (see Fujiwara, 2007 and chapter 2 in this volume).
- Following the Japanese 'Cool Earth 50' proposal of May 2007, the US launched on 31 May 2007 the idea of a 'major emitters' initiative' – later to become major economies meetings (MEM), crucially composed of all large emitters of GHGs in both developed and developing countries, with the aim of developing and contributing to a post-Kyoto framework on energy security and climate change by

[8] Involving the eight member countries of the G8 plus five developing countries – Brazil, China, India, Mexico and South Africa.

the end of 2008. The discussion focused on five key areas: i) low-carbon fossil power generation, ii) transport, iii) land use, iv) market penetration of technologies and v) energy efficiency and finance. Hence, the MEM initiative takes a similar approach as the Gleneagles Plan of Action. See also chapter 2.

- There is an increasing political momentum behind 'global sectoral approaches'. Some industries are concentrated in such a way that a small number of companies represent a large share of the world market and a significant share of the global GHG emissions. While there are different models of approaches across sectors and different models are only emerging, they all attempt to induce changes to technologies through either development of new and breakthrough technologies or accelerated deployment of existing technologies, essentially by means of cooperation between the firms. Global sectoral industry approaches may potentially improve data collection on the state of the sector, e.g. emissions (actual and projected), applied technologies, technology benchmarks and best-practice, identification and spread of best-practice and the development and diffusion of technology (Egenhofer & Fujiwara, 2007).

## 6. Towards an international framework for technology

Many of the technology initiatives can be described as so-called '*technology-oriented agreements*' (*TOAs*). Such initiatives focus on the failures in the market for technological innovation, but they will operate best in conjunction with appropriate emissions-reduction policies, particularly market-based ones. This complementarity could be mutually reinforcing: as emissions-reduction policies spur the uptake of new technologies and increase the profitability of innovation, TOAs will spur additional innovation to lower the costs of mitigation and improve the social and political acceptability of emissions targets. TOAs could be negotiated separately, linked together or incorporated into the climate policy framework in a PAMs approach. More modest TOAs have the advantage of being able to be negotiated and implemented by a smaller set of countries, even outside of the UNFCCC.

De Conink et al. (2007) survey the range of existing TOAs and identify four broad types:

1) Knowledge-sharing and coordination;
2) research, development and demonstration (RD&D);

3) technology transfer; and
4) technology deployment mandates, standards and incentives.

To date, most existing TOAs related to climate change fall into the first category or knowledge-sharing. These initiatives include the Carbon Sequestration Leadership Forum (CSLF), Methane to Markets Partnership or the International Partnership for the Hydrogen Economy, and most recently, the Asia-Pacific Partnership on Clean Development and Climate Change (APP) or existing global sectoral approaches. Agreements of this type offer relatively low-cost means to exchange information, promote common standards and facilitate innovation, but their effectiveness is limited by the voluntary, non-binding nature of the frameworks.

To date, RD&D agreements, i.e. category 2, have primarily focused on fundamental research and demonstration projects, where expenses may be costly and where the technology is too far from commercialisation for intellectual property rights to be a concern. Examples of these cost-sharing arrangements include the ITER fusion reactor, the Solvent Refined Coal II Demonstration Project (SRC-II) and some research conducted by the International Energy Agency (IEA). Technology cooperation and RD&D agreements will most likely reach their limits, however, when the technology moves from the pre-competitive to the post-competitive stage.

Technology transfer, i.e. category 3, has been attempted within the UNFCCC. One of the shortcomings is that most advanced technologies consist of integrated products, which depend to large extent on know-how. Thus, simply transferring patents does not help.

Category 4 of technology mandates is best exemplified by the International Convention for the Prevention of Pollution from Ships (MARPOL). In the climate arena, one could imagine (and some have proposed) harmonised standards for renewable energy, building codes, energy efficiency or requirements for carbon capture and sequestration. These kinds of agreements fall more into 'market-pull' strategies. While they have bigger potential for generating emissions reductions, by virtue of their sector-by-sector nature, they raise cost-effectiveness questions in comparison to broad-based methods. On the other hand, in an international framework with incomplete participation, performance-based standards can have lesser effects on competitiveness than emissions price policies, since standards do not impose the additional direct costs of emissions, resulting in smaller product price increases (Bernard et al., forthcoming; Fischer & Fox, 2004).

The use of TOAs as an effective substitute for an emissions-based approach is limited to the category of standards, mandates or substantial incentives as they would need to be applied on a sector-by-sector, if not technology-by-technology basis. This approach may make the most sense in certain specific settings: for highly trade-sensitive sectors that make agreement on targets and timetables difficult; for sectors not otherwise covered by emissions trading programmes; for sectors that can benefit from international coordination; and for situations where significant ancillary benefits are foreseen. For a comprehensive programme of reducing global emissions, TOAs are best viewed as playing a strong supporting role, alongside a well-designed emissions-reduction policy with long-term targets as the main instrument and driver.

There have been a number of additional practical ideas:

1. A Global Energy Technology Strategy with international cooperation with public and private sectors (along the lines of the APP) can have merits in developing and diffusing technologies, but it ultimately will need to be complemented by incentives (e.g. regulation or carbon price) to provide robust encouragement for the uptake of technologies, as other technology-oriented agreements;
2. The proposal for the Consultative Group on Climate Innovation is a somewhat more concrete version of the above strategy (e.g. Milford, (2007a, 2007b). This approach goes beyond research and focuses on product development, targeted analysis, finance and cooperative policy development. The EU's proposal for an International Energy Efficiency Platform or the idea for a Sustainable Buildings Network point in a similar direction. The new Strategic Energy Technology Plan (SET-Plan) proposed by the European Commission (European Commission, 2007) can be seen as an EU variation of the same approach. Many countries have undertaken technology-mapping exercises.
3. Sectoral approaches that focus on the diffusion of best-available technologies within energy-using sectors.
    a. Trade-sensitive sectors can potentially reach more stringent targets with technology standards than with emissions pricing alone. These standards could build on existing industry-led efforts, such as in cement, aluminium and steel.
    b. Coordinated technology performance standards could help boost conservation by sectors not otherwise covered by

emissions trading programmes (e.g. vehicles or end-use energy demand).

c. Other sectors can benefit a great deal from better international coordination (e.g. building codes, appliance standards, regulation of vessels for international transportation).

d. The CDM could also be a vehicle for sector-based projects for technology transfer in developing countries, particularly if the terms of project eligibility and administrative efficiency are improved.

4. There have been a number of strategies to accelerate technology transfer.

   a. One idea has been the establishment of a Multilateral Technology Acquisition Fund to buy out intellectual property rights (IPRs) and make privately-owned climate-friendly technologies available for deployment in developing countries. This strategy does not directly address the issue of technology development, but would have implications for it: the expected acquisition price helps determine the return on investment to the innovators of that technology.

   b. Removing trade and investment barriers to climate change technology transfer. There are numerous trade barriers in developing countries and industrialised countries that constrain exports of 'greenhouse gas intensity-reducing technologies' in all directions but notably North-South, South-North and South-South (see Brewer, 2007).

   c. Finance is another key element for technology transfer. While a part could be financed by carbon markets, they are probably not sufficient, certainly in the short-term, in which case additional revenues are needed. Some proposals for a global carbon tax or safety valve price for emissions trading suggest that the revenues could be used to encourage a low-carbon transition in developing countries.

   d. The CDM or other post-2012 mechanisms could be strengthened as a vehicle for technology transfer. There is potential to extend such mechanisms towards a broader range of projects, including forestry, bio-energy or adaptation projects or a programmatic or sectoral crediting mechanism.

## References

Barrett, S. (2002), 'Towards a Better Climate Treaty', *World Economics*, 3, 2.

Barrett, S. (2003), *Environment and statecraft*, Oxford: Oxford University Press.

Bernard, A., C. Fischer, and A. Fox, Is There a Rationale for Output-Based Rebating of Environmental Levies? *Resource & Energy Economics*, 29(2), 83-101

Brewer, T.L. (2007), "Climate Change Technology Transfer: International Trade and Investment Policy Issues in the G8+5 Countries", paper prepared for the G8+5 Climate Change Dialogue (available at www.usclimatechange.com).

de Coninck, H., C. Fischer, R.G. Newell, and T. Ueno (2007), *International Technology-Oriented Agreements to Address Climate Change*. RFF Discussion Paper 06-50. Washington, DC: Resources for the Future.

Edmonds, J. (2003), *Toward the development of a Global Energy Technology Strategy to address climate change.* Paper prepared for a strategic roundtable on global energy scenarios of the World Gas Conference, 2 June.

Egenhofer, C, L van Schaik and D Cornland (2005), *Improving the Clean Development Mechanism*. ECP Report No. 1.

Egenhofer, C and N Fujiwara (2007), *Testing global sectoral industry approaches to address climate change.* CEPS Task Force Interim Report, Centre for European Policy Studies, Brussels, December

European Commission (2007), "A European Strategic Energy Technology Plan (SET-Plan) – Towards a low-carbon future", Communication from the Commission to the Council, the European Parliament, the European Economic and Social Committee and the Committee of the Regions, COM(2007)723 final, 22 November.

Fischer, C., and A.K. Fox, forthcoming, "Output-Based Allocation of Emissions Permits for Mitigating Tax and Trade Interactions". *Land Economics.*

Fischer, C., and R. Newell (2004), Environmental and Technology Policies for Climate Change, Discussion Paper 04-05-REV. Washington, DC: Resources for the Future.

Fujiwara, N. (2007), *The Asia-Pacific Partnership on Clean Development and Energy: What it is and what it is not*, CEPS Policy Brief No. 144, Centre for European Policy Studies, Brussels, October.

Galeotti, M. and C. Carraro (2003), "Traditional environmental instruments, Kyoto mechanisms and the role of technical change", in C. Carraro and C. Egenhofer, *Firms Governments and Climate Policy – Incentive-based Policies for Long-term Climate Change*, Cheltenham, UK: Edward Elgar.

Haites E., M. Duan and S. Seres (2006), "Technology transfer by CDM projects", *Climate Policy*, 6(3), pp. 327-344.

Hoffert, M.I., K. Caldeira, G. Enford, D.R. Criswell, C. Green, H. Herzog, A.K. Jain, H.S. Kheshigi, K.S. Lackner, J.S. Lewis, H.D. Lightfoot, W. Manheimer, J.C. Mankins, M.E. Mauel, L.J. Perkins, M.E. Schlesinger, T. Volk and T.M.L. Wigley (2002), "Advanced Technology Paths to Global Climate Stability: Energy for a Greenhouse Planet", *Science*, Vol. 298, 1 November, pp. 981-987.

Humphreys, K. (2001), *The nation's energy future: the role of renewable energy and energy efficiency*. Ken Humphreys' testimony to the US House Committee on Science, 28 February 2001.

Jaffe, A.B., R.G. Newell, and R.N. Stavins (2003), "Technological Change and the Environment". In *Handbook of Environmental Economics* (Vol. 1). Edited by K.-G. Mäler and J. Vincent. Amsterdam: North-Holland/Elsevier, pp. 461–516.

Jaffe, A.B., R.G. Newell and R.N. Stavins (2005), "A Tale of Two Market Failures – Technology and Environmental Policy", *Ecological Economics* 54: 164–174.

Lempert, R.J., S.W. Popper, S.A. Resetar and S.L. Hart (2002), *Capital Cycles and the Timing of Climate Change Policy*, study by the PEW Center on Global Climate Change, Arlington, VA (available at http://www.pewclimate.org).

Milford, L. (2007a), *Massive Climate Technology Innovation – A New Geometry of Complementary Strategies Post-2012*, Statement Paper presented at the Ministerial Meeting of the Gleneagles Dialogue on Climate Change, Clean Energy and Sustainable Development, 9-11 September.

Milford, L. (2007b), *Consultative Group on Climate Innovation – A Proposed Complementary Technology Track for the Post-2012 Period,* paper presented at the Road to Copenhagen Conference on Leadership, Sustainable Development and Climate Change, Brussels, Belgium, 23 November.

Pacala, S. and R. Socolow (2004), "Stablization Wedges: Solving the Climate Problem for the Next 50 Years with Current Technologies", *Science,* Vol. 305, 13 August, pp. 968-972.

Philibert, C. (2003), *Technology Innovation, Development and Diffusion,* OECD and IEA Information Paper: COM/ENV/EPOC/IEA/SLT(2003)4, Paris: IEA.

Sandén, B. and C. Azar (2005), "Near-Term Technology Policies for Long-Term Climate Targets – Economy-Wide versus Technology Specific Approaches", *Energy Policy* 33: 1557–1576.

Tinbergen, J. (1956), *Economic Policy: Principles and Design,* Amsterdam: North-Holland.

World Bank (2006), *An Investment Framework for Clean Energy and Development: A Progress Report,* Vice Presidency for Sustainable Development, Washington, D.C., 1 September.

World Bank (2007), *Clean Energy and Development Investment Framework: The World Bank Group Action Plan,* Washington, D.C., 28 March.

# 10. The Technology Agenda for International Climate Change Policy: A Taxonomy for Structuring Analyses and Negotiations

*Thomas L. Brewer*[*]

## 1. The nature and uses of a taxonomy

The taxonomy presented in this paper focuses on the needs of analysts and negotiators who are involved in negotiations concerning a post-2012 international climate regime.

In practical terms, a taxonomy can be used as a checklist of concepts and issues to be considered in particular negotiating circumstances. More generally, a taxonomy can also serve other related functions, such as structuring an analysis, identifying gaps in an analysis and expanding the scope of an analysis. Furthermore, it can stimulate new ideas as well as codify existing knowledge. Ultimately, however, it can only facilitate analysis; by itself, it is neither a representation of reality nor a set of guidelines for negotiations. The inherent nature of a taxonomy is that much of it is already familiar, since it is developed partly inductively on the basis of previous work, and much of it is cryptic presentation of categories, since it condenses many ideas and much information into short lists and definitions.[1]

---

[*] Associate Professor, McDonough School of Business, Georgetown University, Washington, D.C.

[1] This paper largely ignores semantic issues about the differences between a 'taxonomy' and an 'analytical framework', a 'conceptual model', an 'empirical model' and a 'paradigm' on grounds that they are too academic. Briefly, a 'taxonomy' is a structured list of concepts; an 'analytical framework' or a

A key challenge in developing a taxonomy on any topic is to avoid lapsing into an analytically vacuous exercise that yields a list of ambiguous terms of undemonstrated significance. I have tried to avoid these pitfalls by drawing upon other work on technology, so that the taxonomy is directly responsive to the stated needs of those interested in the technology dimension of climate change. I have also related the taxonomy to others' discussions of the post-2012 climate regime, so that it is grounded in a specific situation. Finally, I have used tangible examples, in order to clarify the meaning and relevance of the terms.

## 2. Elements of the taxonomy

The taxonomy is in four parts, each of which is focused on a particular element of a decision-making process. Together, they encompass a mixture of descriptive, explanatory, evaluative and prescriptive tasks. Although the elements are typically addressed in interactive and iterative decision-making processes in the 'real world', they can be separately identified as follows:

1. *Describing technologies*: What are their key features? How are they similar and different?
2. *Identifying the context of technologies*: How do economic and political factors affect the development and use of technologies?
3. *Evaluating technologies*: What goals and criteria are relevant for evaluating technologies?
4. *Pursuing pathways into the future*: How can the development and use of mitigating or adaptive technologies be facilitated? How can detrimental technologies be constrained?

These elements are discussed separately in this section of the paper and then expressed in the form of an analytical tree in the concluding section.

---

'conceptual model' links concepts verbally; an 'empirical model' links variables mathematically on the basis of systematic empirical data; and a 'paradigm' is a widely accepted understanding (in verbal and/or mathematical form) of a particular phenomenon. For a discussion of paradigms of international climate change technology transfer, see Brewer (2007).

## 2.1 *Describing technologies*

A *generic definition of technology* is that it refers to know-how, whether explicit or tacit, concerning products, production processes or managerial processes. The *form* of technology can be *explicit* or *tacit*. An example of an explicit form is a patent on a product or production process, while knowledge that is internalised in personnel and embedded in organisational processes, such as how to organise an international joint venture for R&D, is an example of a tacit form (Cantwell, 2001).

The *location* of technology can be *production processes* (e.g. how to set up a manufacturing process for a specific item such as a computer chip), or *managerial processes* (e.g. the application of accounting rules to GHG emissions credits) or *products* (e.g. thermostat). The products can be *goods* (e.g. computers) or *services* (e.g. software development).

Another distinction is also sometimes made between *hard* and *soft* technologies. The former refers to tangible things like machines or computers, while the latter refers to knowledge.

There are numerous lists of *industries and products (both goods and services)* related to climate change. Among them are the following:

- In a European Commission (2003) 'Environmental Technologies Action Plan' document, there is a list developed by an advisory group. It includes 51 categories organised in a matrix based on two dimensions. One dimension is the industry sector (e.g. energy supply); and the other consists of energy efficiency/renewables/carbon sequestration/hydrogen & fuel cells. See Table 6.1.
- Pacala & Socolow (2005) identify 15 'wedges' based on the potential contribution of 1 gt$CO_2$e reduction per year by 2054, each one quantified according to the effort needed (e.g. introduce carbon capture and storage at baseload coal-fired power plants of 800 gigawatts).
- Stern (2007, p. 259) provides a list of nine types of technologies that could reduce carbon emissions in the energy sector: efficiency, carbon capture and storage, nuclear, biofuel, dCHP, solar, wind and hydro.
- The US Climate Change Technology Program (2006) itemises hundreds of technologies, which are listed in 'current portfolios' and 'future research directions'. They are organised according to end-use/infrastructure (e.g. transportation), energy supply (e.g. hydrogen), carbon capture-storage (e.g. geologic storage), non-$CO_2$

GHGs (e.g. methane from landfills) and measuring & monitoring capabilities (e.g. oceanic $CO_2$ sequestration).

- A World Bank study (2007a) of trade issues related to climate change includes 43 climate-friendly *goods* identified by the code numbers used in the Harmonised System (HS) for international trade.

Note that most of the items above rely at least in part on *industry sectors* as a basis for the classification of technologies. If the international negotiations on a post-2012 climate regime include a sectoral approach, among others, then the specific sectors used in the negotiations would be a natural basis for defining technologies.[2]

There are *related definitions.* For instance, 'clean and efficient energy technology' has been defined to include technologies that result in reduced emissions of GHGs.[3]

---

[2] When identifying technologies in terms of standardised product or industry classification schemes, such as the Harmonised System (HS) of the World Customs Union, or the International Standard Industrial Classification system or the UN Product Code, there are a variety of technical issues, which can be important in negotiations. These and related issues have been raised recently, especially in regard to international technology transfer and thus trade issues (see Howse, 2006; OECD, 2006; Sugathan, 2006; World Bank, 2007). Furthermore, at the WTO, classification issues are different for goods and services because the General Agreement on Tariffs and Trade (GATT) and the General Agreement on Trade in Services (GATS) use different product classification schemes (Brewer, 2007).

[3] The complete definition in the proposed International Climate Cooperation Re-engagement Act of 2007, which is part of a larger pending US energy bill, reads as follows [italics added]: 'The term "clean and efficient energy technology" means an energy supply or end-use technology—(A) *such as*— (i) solar technology; (ii) wind technology; (iii) geothermal technology; (iv) hydroelectric technology; and (v) carbon capture technology; *and* (B) that, over its life cycle and compared to a similar technology already in commercial use — (i) is reliable, affordable, economically viable, socially acceptable, and compatible with the needs and norms of the country involved; (ii) results in — (I) *reduced emissions of greenhouse gases; or* (II) *increased geological sequestration; and* (iii) may — (I) substantially lower emissions of air pollutants; *or* (II) generate substantially smaller or less hazardous quantities of solid or liquid waste.'

*Table 6.1 Identification of key technologies for the reduction of greenhouse gases*

| | Energy supply | Energy demand-households + services | Energy demand industry | Transport | Agriculture | Waste |
|---|---|---|---|---|---|---|
| **Energy efficiency** | -Advanced macro CHP<br>-Micro CHP<br>-Coal bed methane<br>-Ultra-high efficiency combined cycle gas turbines<br>-High efficient clean coal technology | -Building fabric<br>-Integrated building design<br>-Controls & building energy management systems<br>-Heating & cooling & ventilation equipment<br>-Energy efficiency equipment-office and domestic equipment<br>-Lighting | -Alternative equipment<br>-Combustion technologies<br>-Low temperature processing materials<br>-Process control<br>-Separation technologies<br>-Waste heat recovery | -Improvement internal combustion engine (diesel & gasoline)<br>-Hybrid vehicles<br>-Aeronautic technology<br>-Traffic management systems | -Diet composition for reduced enteric fermentation | -Waste treatment technologies<br>-Recycling/recovery (including eco-design) |
| **Low carbon-technologies**<br><br>**Renewables** | -Direct solar (Photovoltaic, Solar thermal power stations)<br>-Wind onshore/offshore<br>-Biomass-electricity generation<br>-Geothermal<br>-Tidal wave<br>-Small hydro | -Biomass-local heat generation<br>-Passive solar systems | -Biomass-process heat | -Bio-fuels-transport | -Production of biomass | -Capture of bio-gas |
| **Low carbon technologies-$CO_2$ sequestration** | -$CO_2$ capture and storage (various options) | - | - | - | -Biological carbon sequestration | - |
| **Hydrogen & fuel cells** | -Production of hydrogen from renewable energy sources (including options such as photo electrolysis), fossil fuels with $CO_2$ sequestration | -Fuel cells-domestic CHP | -Fuel cells- industrial | -Hydrogen internal combustion engine<br>-Fuel cells-transport<br>-Hydrogen storage<br>-Hydrogen infrastructure | -Production of biomass for hydrogen productions | |

*Source*: European Commission (2003), pp. 12-13.

## 2.2 *Identifying the context of technologies*

Differences among countries in their economic and political institutions, as well as their official status according to UNFCCC and Kyoto Protocol Annexes need to be taken into account. These differences include, in particular, countries' varying technological *needs* (UNFCCC, 2007) and *institutional capacities* for absorbing new technologies (Mytelka, 2007).

Key issues about the economic context of technologies concern *market failures*. There are two sets of problems. First, there needs to be a price on GHG emissions to internalise the costs and create incentives for technological change. Second, other specific technology market problems need to be overcome to avoid underinvestment in climate technologies. These include (Fischer & Egenhofer, 2007, pp. 1-2):

- *Knowledge spillovers,* i.e. innovating firms cannot capture all the benefits of their innovations because some of the new knowledge spills over to their competitors and other firms;
- *Adoption spillovers,* i.e. benefits accrue as a given technology becomes more widely adopted (sometimes called 'network externalities'); and
- *Incomplete information,* i.e. information about future government climate change policies and thus carbon prices, as well as information about the effectiveness of technologies.

As for the political context, the *level* of the institutional forum in which the issues are being considered and/or will be implemented needs to be taken into account. The levels are multilateral (e.g. UNFCCC), plurilateral (e.g. WTO Government Procurement agreement), inter-regional (e.g. EU-Mercosur), regional (e.g. NAFTA), national and subnational.

*International* technology transfer involves distinctive issues that arise from a) differences in the economic, political-legal and cultural features of societies and b) barriers to transfers as a result of government international trade, investment and technology transfer policies (Brewer, 2007; IEA, 2001; UNFCCC, 2007).

*Private technology transfers,* including by multinational firms (Barton, 2007; Brewer, 2007; Cantwell, 2003), are especially important in international technology transfers. Private sector modes of transfer include: trade, investment, licensing and human mobility through employee international transfers and individual migration.

### 2.3 *Evaluating technologies*

There are at least six *evaluative criteria* for assessing technologies:

1) *Environmental impact.* What are the potential reductions of GHG concentrations and emissions? Which technologies have the potential for the largest reductions of GHG emissions?
2) *Technical feasibility and efficiency.* Which technologies are the most feasible and/or efficient in engineering terms?
3) *Economic cost-benefit relationships.* What are the relative monetised cost-benefit relationships among technologies?
4) *Political effects.* What will be the distribution of costs and benefits of adopting the technology? Who pays and who benefits from a technology? Who supports it/opposes it?
5) *Social effects.* What are the beneficial and detrimental non-economic impacts of adopting the technology? Which communities or other groups experience renewal and which experience decline or dislocation?
6) *Ethical issues.* How do the political and economic and social impacts of the adoption of a technology vary across groups in terms of their responsibility for creating the problem? Do relatively wealthy or poor groups pay or gain?

This paper emphasises *mitigation* technologies, but of course there are also numerous technologies for a*daptation.*

*Climate-friendly* technology is a relative concept. Further, technologies can become more or less climate-friendly over time, as technologies change. "Energy efficiency", for instance, is a "relative and evolving concept" (Sugathan, 2006, p. 8). For instance, as progress is made in home appliances to make them more energy efficient, earlier models become less energy efficient in relative terms.

The specific goals of a technology can be described in terms of its:

- *Scope* – Are the goals global, regional, national or sub-national?
- *Time frame* – Are the goals short-term (within a year), medium-term (1-10 years) or long-term (more than a decade)?
- *Relationship to other goals* – Are particular technologies also relevant to energy security and/or sustainable economic development? Are specific technologies mutually reinforcing, compatible or conflicting?

Specific technologies can constitute *threat wedges* (Wellington, et al., 2007, p. 6), that is, technologies that are adopted for objectives other than climate change mitigation and that would increase carbon emissions. Three such technologies have received much attention in the context of energy security concerns, namely the production of synthetic fuels from coal, the extraction of oil from tar sands and the production of oil from crushed rocks (oil shale).

### 2.4 *Pursuing pathways into the future*

There are numerous *stages in the technology innovation process*, including basic R&D, applied R&D, demonstration, commercialisation and diffusion (Stern, 2007, pp. 395-399). Government policies often focus on the first three stages (sometimes with business partnerships), but there are also numerous ways for governments to facilitate commercialisation and diffusion - including subsidies for production, investments or purchases, market share mandates, performance standards, liability limitations, government procurement requirements, consumer information regulations, infrastructure development and international technology transfer support (Newell, 2007, p. 3).

*Portfolios* of technologies are useful ways to address the problem of *a priori* uncertainties about the cost-effectiveness of individual technologies (Stern, 2007, pp. 273, 407-409, 418-419).

Programmes for climate change mitigation or adaptation can *interact* with other climate change programmes. The proceeds from cap-and-trade allowance auctions or revenues from climate taxes can be used to support technological research and/or other stages of the innovation process. Issues that may arise from such uses of funds include: How much money is available? What are the technological objectives? What are the constraints on the use of the funds?

Estimating and supporting the *scale of the technological effort* needed to address climate change issues is problematic (Wellington et al., 2007). Stern (2007, p. 394) estimates that a doubling of present annual world-wide government expenditures on *energy R&D* to $20 billion is needed (also see UNFCCC, 2007 and World Bank, 2007b). An example of large-scale *diffusion of many technologies at the project level* is the development on an island near Shanghai, China, of an entirely new 'eco-city' that is being built in a previously rural area (Stern, 2007, p. 437).

## 3. Analytical tree

The analytical tree depicted below highlights key nodes and choices when applying the taxonomy. This is only a partial synopsis; the entire analytical process in reality is of course highly iterative and interactive. In reality, the elements depicted here are typically addressed repeatedly and in a variety of sequences, when applied to a particular situation.

*Figure 6.1 Analytical Decision Tree*

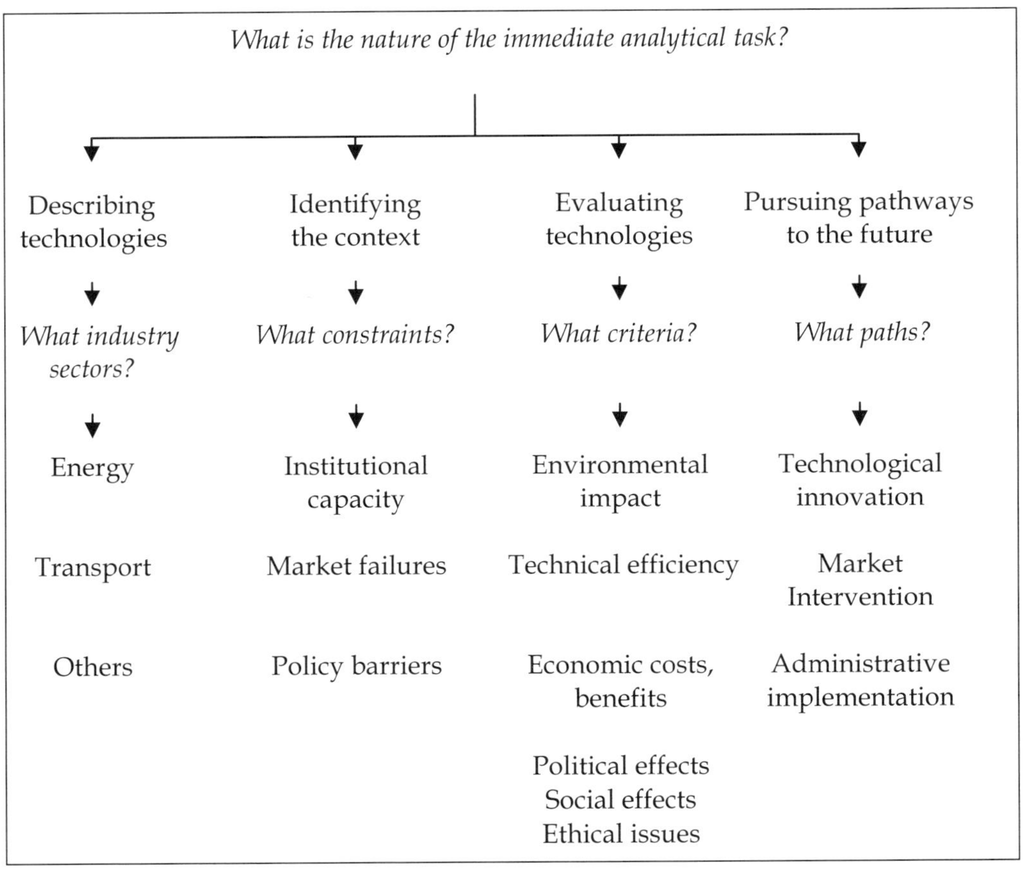

## Annotated References and Suggestions for Further Reading


Barton, John H. (2007), *New Trends in Technology Transfer: Implications for National and International Policy*, Issue Paper No. 18, International Centre for Trade and Sustainable Development, Geneva (www.ictsd.org). *Importance of private channels, especially multinational firms, in technology transfer.*

Brewer, Thomas L. (2007), "Climate Change Technology Transfer: International Trade and Investment Policy Issues in the G8+5 Countries", paper prepared for the G8+5 Climate Change Dialogue (www.usclimatechange.com). *Barriers to technology transfer in the form of trade and foreign direct investment policies.*

Cantwell, John (2001), "Innovation and Information Technology in the MNE", in Alan M. Rugman and Thomas L. Brewer (eds), *The Oxford Handbook of International Business*, Oxford: Oxford University Press, pp. 431-456. *Useful introduction to concepts and literature of international business on technology transfer.*

Edmonds, Jae, Joseph M. Roop and Michael J. Scott (2000), *Technology and the Economics of Climate Change Policy*, Pew Center on Global Climate Change (www.pewclimate.org).

European Commission (2003), Environmental Technologies Action Plan.

Fischer, Carolyn and Christian Egenhofer (2007), "The Critical Role of Technology for International Climate Change Policy", Centre for European Policy Studies, Brussels (www.ceps.eu). *Wide-ranging analysis of the issues.*

International Energy Agency (IEA) (2001), *Technology without Borders: Case Studies of Successful Technology Transfer*, IEA, Paris (www.iea.org). *Few short studies of factors contributing to success in technology transfers.*

Justus, Debra and Cedric Philibert (2005), *International Energy Technology Collaboration and Climate Change Mitigation: Synthesis Report*, OECD, Paris (www.oecd.org). *Important summary of issues.*

Mytelka, Lynn (2007), *Technology Transfer Issues in Environmental Goods and Services: An Illustrative Analysis of Sectors Relevant to Air-pollution and Renewable Energy*, International Centre for Trade and Sustainable Development, Geneva.

Newell, Richard G. (2007a), "Climate Technology Policy", Resources for the Future, Washington, D.C. (www.rff.org). *Brief introduction to central concepts and issues.*

Pacala, Stephen W. and Robert H. Socolow (2004), "Stabilization Wedges: Solving the Climate Problem for the Next 50 Years with Current Technologies", *Science*, 305, 13 August 13, pp. 968-972. *First publication of important list of technological wedges.*

Stern, Nicholas (2007), *The Economics of Climate Change: The Stern Review*, Cambridge: Cambridge University Press. *Includes several sections on technology issues and cites many previous studies including work of the UNFCCC.*

Sugathan, Mahesh (2006), "Climate Change Benefits from Liberalisation of Environmental Goods and Services", in Erwin Rose and Moustapha Kamal Gueye (eds), *Linking Trade, Climate Change and Energy*, International Centre for Trade and Sustainable Development, Geneva (www.ictsd.ch). *Focused on trade liberalisation issues.*

Ueno, Takahiro (2006), *Reengineering the Climate Regime: Design and Process Principles of International Technology Cooperation for Climate Change Mitigation*, Resources for the Future, Washington, D.C. (www.rff.org). *Analysis of eight international technology cooperation projects and agreements.*

United Nations Framework Convention on Climate Change (UNFCCC) (2007), Background Paper on Analysis of Existing and Planned Investment and Financial Flows Relevant to the Development of Effective and Appropriate International Response to Climate Change (www.unfccc.org). *Extensive, detailed analysis, by region and country, of current and expressed needs for investments, financing and technologies for climate change mitigation and adaptation.*

US Climate Change Technology Program (2007), *Strategic Plan* (www.climatetechnology.gov). *Lengthy presentation of US government programs and plans.*

Wellington, Fred, Rob Bradley, Britt Childs, Clay Rigdon and Jonathan Pershing (2007), *Scaling Up: Global Technology Deployment to Stablize Emissions*, World Resources Institute and Goldman Sachs Center for Environmental Markets, Washington, D.C. and New York (www.wri.org). *Includes data on private investment flows as well as specific government policy issues associated with types of energy technologies.*

World Bank (2007a), *Warming Up to Trade: Harnessing International Trade to Support Climate Change Objectives*, Washington, D.C. (prepared by Muthukumara Mani and others) (www.worldbank.org). *Major study of international trade issues and- climate friendly technology transfers.*

World Bank (2007b), *Clean Energy and Development: Toward an Investment Framework*, Washington, D.C. (www.worldbank.org). *Wide-ranging analysis of energy issues, including especially climate change issues, and related investment needs in developing countries.*

# 11. WHY INTERNATIONAL TRANSPORT NEEDS A SECTORAL APPROACH

*MARKUS ÅHMAN**

## 1. Introduction

The increased mobility of people and goods has been a critical factor in the phenomenal economic growth the world has experienced over the last century. Globalisation and growing international trade have had a deep impact on the world economy and helped many people across the world attain a higher standard of living. Some would even argue that the rapid development of international transport has made the world a better place.

However, it is also evident that international transport poses a serious threat to the world's climate system. It is one of the fastest growing sources of greenhouse gas emissions, and there is currently very little being done to curb this trend.

This chapter discusses the background to the current regulation of international aviation and maritime shipping and gives an overview of the options to address emissions. It then offers an argument for why a sectoral approach is the most promising policy avenue for including these modes of transport in a post-2012 climate agreement.

* IVL Swedish Environmental Research Institute, the Clipore Research Team and member of the ECP Steering Group. The author would like to acknowledge David Mjureke (Swedish Ministry of the Environment), Åsa Löfgren (Göteborg University), Nardia Haigh (University of Queensland) and Asbjørn Torvanger (CICERO) for helpful comments.

## 2. Should we care about these emissions?

Yes. Emissions from international aviation and shipping[1] constitute a small, but rapidly increasing percentage of global greenhouse gas emissions, accounting for between 3-6% each today. In the period 1990-2004, these emissions, commonly referred to as emissions from 'bunker fuels', have grown 34% and 43%, respectively. If growth continues along its current path, emissions from international aviation and shipping could account for about 15% of global emissions by 2050. Forecasts for the sector suggest that EU international aviation emissions will increase by 150% between 1990 and 2012, which would offset more than a quarter of the reductions required by the Community's target under the Kyoto Protocol.

To understand the problems and potential ways forward, it is useful to look at the background to the current regulation since it has a slightly different history and dynamic than that of land-based activities.

## 3. How did we end up here, and what is being done about it?

To date, relatively little has been achieved that could slow the trend of growing emissions from international aviation and shipping. In fact, concrete results that could lead to significant mitigation of emissions from bunker fuels have been largely absent from the UN process. Emissions from bunker fuels have been part of the UNFCCC negotiations since before the Convention entered into force, but bunker fuels are not included in the targets specified in the Kyoto Protocol. For a long time, discussions have revolved around how to calculate and report emissions rather than around concrete action or commitments to reduce them. Since 2005, parties have not been able to agree on any conclusions concerning bunker fuels.

In parallel to the UNFCCC, the International Civil Aviation Organisation (ICAO) and International Maritime Organisation (IMO) have carried out work.

ICAO has had climate change on its agenda for a number of years and has set up a working group charged with developing a proposal on how to limit the climate impact from aviation, although no firm reporting date has been set for the group. The current legislation on international

---

[1] For the remainder of this chapter, the term 'shipping' refers to maritime shipping by boat.

aviation is founded on the concepts in the Convention on International Civil Aviation, also known as the Chicago Convention that entered into force in 1947. The preamble to the convention states that civil aviation should be established on the basis of "equal opportunity", a priority that still dominates talks on regulation of international aviation. ICAO has stated that it is positively disposed towards emissions trading in principle, but at the same time it adopted a resolution at its general assembly in 2007 requiring any regional trading scheme to have bilateral agreements with any third party affected by the system. In practice this makes the EU initiative to include international aviation in the EU ETS impossible.

The division of the sea and its resources is regulated through the United Nations Convention on the Law of the Sea (UNCLOS). The current formulation restricts implementation of local and regional rules for international shipping. Policies to address emissions, unless associated with some service,[2] would have to be voluntary according to the Convention. IMO has set up a work plan on greenhouse gas reductions, scheduled to finish in 2010. Member states will then have to decide if the IMO should establish a mandatory regime for the control of greenhouse gas emissions from shipping. To date, the principal result of the work plan is the development of a $CO_2$ index - a description of $CO_2$ efficiencies of ships including emissions per tonne of cargo, passenger, nautical mile, etc. This is now being introduced in a trial period until 2008.

The striking lack of progress in the work at the international level begs the question: What is it that makes dealing with these emissions so difficult?

## 4. What makes international transport special?

There are three principal factors that can help to explain the failure of international regulation in this field. First, the significance of international transport to the global economy makes the issue sensitive. Cheap transport is a major lubricant to trade and economic growth, and governments are reluctant to introduce measures that could impede their functioning and act as barriers to trade. Although barriers to trade may reduce emissions,

---

[2] There are currently local fees for navigation, for instance, the fairway fees in Sweden. These can be applied since they are service-based, e.g. maintaining fairways in good condition.

they tend to be inefficient. Trade allows countries to develop comparative advantage in production and therefore, for instance, border tariffs are clearly second best to globally agreed policies aimed directly at reducing greenhouse gas emissions. There are also concerns that the desire to reduce transport emissions could become a pretext for other measures that are essentially protectionist and support inefficient industries. Since such measures could make it considerably more difficult to build the trust necessary for future international cooperation, these concerns should be taken seriously. Furthermore, a key driver behind growth in transport emissions is income; as people get richer, they tend to want to transport more goods and make longer trips. Transport emissions increase both because people travel more, and because as people get richer they tend to travel using more carbon-intensive modes (switching from bus to train, from train to air, from small car to large car, etc.). This link to economic development at the personal level adds emotional resistance to regulating transport.

Second, countries' limited jurisdiction in controlling emissions from activities occurring outside their national borders adds to the intricacies in the negotiations on policies. Existing international agreements place significant restrictions on what countries are required to do to curb emissions, and also on what they are *allowed* to do. For instance, there are legally binding provisions in a number of bilateral air service agreements that make it difficult (in the short and medium term) to introduce taxes that could distort competition between carriers of different nationalities. However, there are limits to these restrictions. For example, Article 2.2 of the Kyoto Protocol states that industrialised countries shall seek to limit emissions from bunker fuels through work in the ICAO and the IMO. If ICAO and IMO do not succeed in this, emissions might be regulated at the regional level, as long as the principles of 'equal opportunity' in the case of aviation and 'innocent passage' for maritime shipping are respected. This would suggest that countries do have the necessary legal authority to set emissions limits and reporting requirements, at least in the medium term.

A third challenge is monitoring, reporting and verification. The fact that ships and airplanes move across national borders and jurisdictions makes monitoring and reporting difficult. This is particularly true for shipping since ocean-going vessels can bunker very large quantities of fuel. An issue in the relations between UNFCCC on the one hand and ICAO/IMO on the other hand is the split in Annex 1/non-Annex 1 countries, which is used under the UNFCCC/Kyoto Protocol but not by

ICAO or IMO. If a future agreement should make non-Annex I countries exempt from commitments, this must be dealt with. It is worth noting that although 75% of ships are registered in non-Annex I countries, a majority are owned by companies in Annex I countries.

The national inventories of greenhouse gas emissions only include emissions occurring within the national territory, while international aviation and shipping are reported separately as so-called 'memo items' by the fuel-selling country. Any effective policy regime would need reliable data on the volume of emissions and a mechanism to allocate these emissions to an entity or party that can be held accountable in an agreement. For example, offshore bunkering, lack of clear definitions and errors in reporting are problems that cause statistical uncertainties in bunker fuel sold in the shipping sector. However, improvements in technology and reporting systems over the last decade strongly suggest that the remaining technical issues related to monitoring and reporting can be solved. For instance, the IMO already has a long-range tracking system that could provide for an activity-based approach to emissions reporting, which could give more precise data than marine bunker sales. Further, ship owners already have accurate data on bunker fuel volumes used. The same is true in the aviation sector; very detailed data exist in aviation companies and could be collected if so decided. There are no technical barriers to using such data, should reporting be made mandatory. In summary, the absence of global policies and measures seems due to institutional issues and political barriers rather than to technical shortcomings.

In addition to these general factors, there are more specific issues associated with each mode of transport. For instance, the climate impact from emissions from aviation does not only depend on emitted volume, but also on flight altitude and weather conditions. The total climate impact of emissions from aviation is uncertain but it is significantly higher than the impact by $CO_2$ emissions alone. Important emissions are $NO_x$, $SO_2$, soot (particles) and water vapour. Furthermore, it has proven more difficult to apply definitions and methods for reporting in the shipping sector than in the aviation sector. Thus it is important to remember that aviation and shipping are qualitatively different and it is not obvious that they should be dealt with in the same manner.

## 5. Is there a way forward?

Against the backdrop described above, it is easy to be pessimistic. The only major new international or regional regulation currently planned is the inclusion of aviation in the EU ETS. It is scheduled for 2012, but the decision has already led to protests from other parties of the ICAO who claim that it is illegal of the EU to impose such a policy on flights that take off or land outside of Europe. The EU has also stated that maritime shipping might be included in the EU ETS, should the IMO not succeed in reaching an acceptable solution in 2009. To what extent new developments such as the launch of the Bali Roadmap, the election of a new government in Australia and the upcoming presidential election in the US can spawn new policy initiatives on international transport remain to be seen, but so far the prospects appear unchanged in the main.

Nevertheless, there are also positive developments and promising policy avenues. The evolving emissions trading systems across the world and the maturing discussion on how to design such systems is one such development. The increased attention given to the potential in sectoral approaches is another. On the technical level, the growing consensus within the ICAO and the IMO that policies to control emissions are feasible from a reporting and monitoring perspective is a third critical development.

So what are the policy options looking forward?

- The regulatory instruments referred to as 'command and control' are often specific, demanding certain technologies to be implemented or quantified environmental standards to be met by individual polluters. Polluters' compliance is based on monitoring and enforcement. Traditionally, these types of policies are regarded as effective, easy to manage, relatively simple to impose and broadly accepted. However, they are generally not cost-effective and therefore not socially optimal.
- Incentive-based instruments aim to internalise costs related to damages to the environment into the market by putting a price on emissions, thereby providing 'market signals' in the form of a modification of relative prices. Emissions charges or taxes, user charges, product charges or taxes, administrative charges or fees, the emissions trading system (ETS), deposit-refund systems and subsidies are all examples of incentive-based policy instruments. In theory, these types of instruments have all the efficiency properties of

competitive market pricing, which guarantees an efficient allocation of resources in the economy, provided all costs are accounted for.

Command and control approaches have frequently been put forward as the most feasible for international transport. Indeed, they are already used for other environmental problems, for instance by countries requiring double-hulled ships for certain routes. Looking ahead, it will be interesting to follow the discussion on standards on emissions levels and fuel efficiency.

However, the most spectacular policy development during the last decade is the growing political popularity of emissions trading. Right now there are systems in place or being planned in all developed countries.

Most proposals for a post-2012 climate agreement include emissions trading in some form. Many observers would like to see trading of emissions between countries extended to direct trading between companies. The option to include international transport emissions into a global emissions trading system is receiving increasing attention. The attraction of this is threefold:

1) As emissions from bunker fuels are virtually uncontrolled today, almost any form of policy would be an improvement over the current situation.
2) Including bunker fuel emissions under a cap could put an end to the present, seemingly unstoppable, growth.
3) The cost-saving potential of a trading system increases with broader coverage, which speaks in favour of integrating all major emitting sectors.

The realisation of these benefits, however, requires some careful thinking when designing the trading system.

At first glance, a trading system with allocation of emission allowances at the country level may seem like the most straightforward approach. However, the difficulties of assigning responsibility for emissions from international transport to individual countries seem likely to remain. Negotiations on this issue have so far produced little result, and the outlook for the future looks grim. This makes the feasibility of a trading system for bunker fuel emissions based on national allocations questionable.

Furthermore, as discussed in the next chapter of this book, a quick and full integration of transport and industry in the same emissions trading system seems unwise for economic reasons. The large differences in abatement costs and willingness to pay for emissions between industry and transport could trigger significant negative effects on industry. There is a relevant argument that if the cheapest abatement options exist in industry, then those are the ones that should be pursued. But at least in the medium term, the risk of severe economic disruptions in industry seems to outweigh the advantages offered by a fully integrated trading system. It is also unlikely that emissions from bunker fuels would be reduced significantly in an integrated system since most reductions would be carried out in industry.

A potential solution would be to combine an emissions trading system with some form of carbon tax on bunker fuels, thereby offsetting some of the difference in abatement costs. This option has the advantage of controlling transport emissions while protecting industry from radically increased allowance prices. Although the economic benefit of the trading system would be reduced by such a solution as it would mean that different parts of the economy would pay different prices for emissions, the tax could gradually be reduced over time as abatement costs in transport approach those in industry. This said, reaching international agreements on taxes has proven extremely difficult. Thus any policy regime that is founded on such agreements is likely to fail, which basically excludes a global carbon tax as well as an integrated system with industry and international transport from the palette of policy options.

## 6. What about a sectoral approach?

Another possibility would be to use a sectoral approach in the construction of the trading system. In this approach, separate emissions trading systems for industry and transport would be created. A target would be set for each system and the participants would not be able to trade allowances with each other. Considering the important differences between aviation and shipping, it is possible that they should be treated separately. As the formidable task of reaching a post-2012 climate agreement has become increasingly real to policy-makers, the concept of sector-based climate agreements has received growing attention. In fact, it is mentioned explicitly in the Bali Roadmap. There are several characteristics of international transport that make a sectoral approach particularly attractive:

- Simply the fact that reaching a multilateral agreement based on countries has proven so difficult warrants serious consideration of a different approach, such as sectoral trading systems.
- International transport is similar regardless of which country it takes place in. Thus even though countries may differ in many respects, it could be possible to reach an agreement on a target for aviation and shipping.
- Transport is, almost by definition, linked to international competition and cross-border trade, which are two issues where sectoral agreements are suggested to be particularly useful.
- Emissions sources move across borders, which makes assigning responsibility and avoiding leakage even more difficult than for other goods and services. Assigning emissions to activity or operator would simplify this problem.
- The climate impact of aviation emissions is complex and differs from that of emissions at the earth's surface, even if the emitted volumes are identical. This makes trading emissions between sectors problematic.
- International transport, in particular aviation, is dominated by a relatively small number of actors. This could ease the negotiating of an agreement considerably.

An important advantage of a sectoral solution is that we can effectively put a cap on transport emissions. The disadvantage is that the total 'global' costs to reach a given global emissions target are higher than with an integrated ETS, since potentially cheaper emissions reductions in the industrial sector cannot be traded against emissions in the transport sector.

Another advantage is that we can expect a lower impact on allowance price in the industrial ETS, since there will no longer be a demand for industrial allowances from the transport sector. Thus, the impacts from increased allowance price, such as increased operation costs, increased electricity price and risks for closures and carbon leakage, will be reduced.

## 7. Conclusions

Emissions from international aviation and maritime transport constitute a small, but rapidly increasing, percentage of global greenhouse gas emissions. A business-as-usual scenario could put emissions from international aviation and maritime shipping at 15% of global emissions by 2050. To date, very little has been done to curb this trend.

This development, coupled with the growing political popularity of emissions trading, has put the spotlight on the option of an integrated international trading system, which would include international aviation and shipping. The idea is to address emissions from international aviation and shipping and capture the cost savings that a broad emissions trading system offers.

However, the characteristics of bunker fuel emissions, with close links to international trade and competition, the difficulties of allocating emissions on a country level and the complexities in estimating climate impact from emissions make the feasibility and effectiveness of a conventional emissions trading system questionable. Furthermore, a full integration of international transport and industry could cause significant negative effects for industry, without achieving substantial cuts in bunker-fuel emissions. In the medium term at least, these negative effects seem to offset the advantages of an integrated system.

Instead, a sectoral approach seems to be the most promising option if a cap and trade system for bunker fuels is to be negotiated in the medium term. Even though this would, in theory, raise the total cost of reaching a global emissions target as compared to an integrated emissions trading system, it offers several advantages. Most importantly, it seems feasible from a technical and political point of view and would allow the negative effects on industry to be managed, without compromising an effective control of emissions from international aviation and shipping.

# 12. How to Integrate International Aviation and Shipping into a Global Emissions Trading System

*Lars Zetterberg*[*]

## 1. Introduction

Greenhouse gas emissions from international aviation and shipping are growing at an alarming rate and little progress has been achieved in controlling these emissions. Meanwhile, substantial progress is being made in the development of international emissions trading. The EU emissions trading system is the most prominent main example, but we are also seeing the emergence of other national and regional emission trading systems, for example in Norway, Iceland, North Eastern USA, Australia and New Zealand. It is likely that there will be linkages between these systems and we may soon see the first steps of what could become a global emissions trading system. With this development, it is natural to address the possibility of including greenhouse gas emissions from international aviation and shipping into a global cap and trade system. The linking of transport to a global cap and trade system will be a crucial future challenge for the global climate change regime. It is unclear what the extent and distribution of costs and reductions would be on different actors. However, the example of integrating road transport into the EU emissions trading system holds many lessons for international aviation and shipping. The objectives of this paper are to discuss the pros and cons of integrating international aviation and shipping with a global industrial cap and trade system.

* IVL Swedish Environmental Research Institute and member of the Clipore research team.

## 2. What does experience and current research on emissions trading and transport tell us?

Given the emergence of emission trading systems, it is obvious to address the possibility of including international transport emissions into a global emissions trading system (ETS). There are several attractions to such a development:

1. Firstly, greenhouse gas emissions from international aviation and shipping are not controlled today. It is crucial to find a solution to control these bunker-related emissions.
2. Secondly, emissions from international aviation and shipping are growing at an alarming rate. This is partly due to an increased demand for international transport, but can also be explained by the absence of policy instruments. Putting these emissions under an emission trading cap could be an effective way to control them.
3. Thirdly, it is advantageous to broaden an emissions trading system by adding new sectors, countries and gases. In this way, more options for emissions reductions are available, and the total costs for achieving a given emissions target can be further reduced. Because international transport accounts for a considerable share of the carbon emissions globally, economic savings could potentially be achieved if transport were integrated in a global ETS.

In order to understand the consequences of linking international transport to a global industrial ETS, important lessons can be drawn from recent theoretical studies of integrating EU transport with the EU ETS. Several studies have investigated the implications of integrating road transport into the EU ETS, e.g. Holmgren et al. (2006), CE Delft (2006) and Kågeson (2008).

The crucial factor that determines the economic consequences of an integration is the difference in abatement costs between the two systems that are linked, in this case the transport sector and industry. This factor is the main determinant for the total cost savings, the distribution of emissions reductions and the distribution of costs among the participating actors. The larger the difference in abatement costs, the larger is the potential for total cost savings, but also the larger will be the redistribution of reductions and costs.

If we first look at the EU example, it appears that the marginal abatement costs or at least the willingness to pay for emissions are considerably higher in the EU transport sector than in EU industry. This can be seen by comparing current carbon taxation levels in the two sectors. The current taxation on petrol in the EU corresponds to a carbon tax that is 10 to 20 times higher than the carbon taxation in industry, assuming an allowance price of €15. This difference in willingness to pay determines the consequences of including transport in the EU ETS.

In contrast, international transport is not subject to carbon taxes at all. One could therefore argue that carbon abatement costs would be relatively cheap here. However, in aviation and shipping, fuel price is a major part of the operation costs and therefore considerable effort has been invested in increasing fuel efficiency. Since carbon emissions are closely related to energy consumption, these energy efficiency measures have also contributed significantly to reducing carbon emissions. It is therefore unlikely that we can harvest cheap carbon reductions through energy efficiency measures. This is also confirmed by studies that show that significant emissions reductions (more than 20%) in aviation and shipping are costly. So if transport emissions were to be restricted, for instance under an absolute emissions cap, and transport volumes continue to grow, the cheaper emissions reductions would soon be exhausted and abatement costs would then increase significantly.

Even if the scientific evidence so far is incomplete, let us assume that abatement costs for international aviation and shipping are high. With this assumption, the case of linking international transport to a global industrial ETS would be similar to the case of including EU transport in the EU ETS. Both cases represent the linking of two systems with different abatement costs and we can expect similar patterns of consequences. Based on findings in recent theoretical studies of including EU transport in the EU ETS, we can estimate the consequences of integrating international transport into a global ETS.

In order to analyse the consequences of integrating international transport into a global industrial ETS, we need to define a reference case. In this reference case we assume a global ETS without transport and that there is a requirement on international transport to reduce emissions within its own sector, either by implementing an ETS specifically for the transport sector or by means of another policy instrument.

If transport were integrated into the same emissions trading system as industry, the two sectors would compete over the same emissions reductions. Initially, low-cost reductions may exist in international shipping and aviation since these sectors don't have a carbon tax today. When these 'low hanging fruit' have been harvested and international transport volumes had grown, the competition for emissions reductions would increase. As a consequence, the price of allowances would increase.

Assuming that transport has a larger willingness to pay than industry, the transport sector would buy emissions reductions from industry and continue to release carbon to the atmosphere.

In industry, regardless of allocation, the marginal operating costs, including the shadow price of allowances, would increase. The price of electricity would increase in liberalised markets and for some industries this would constitute a double impact (higher price for allowances and electricity). On the other hand, with higher allowance prices, new carbon-efficient technologies that previously were not economically viable, may become profitable and experience a market breakthrough.

Production in carbon-emitting industries would decrease and could lead to closures or reallocation of industries to countries outside the emissions trading system (carbon leakage). This risk would be lower the more economies that are included in the ETS globally.

In the international transport sector, compliance costs would decrease as compared to a situation where international transport has to reduce emissions within the sector. This is likely to lead to continued increasing emissions. Ongoing carbon reduction programmes, with relatively high reduction costs, such as low-carbon fuel chains and $CO_2$-efficient vehicles, may become unprofitable.

For emissions trading in general, increasing the number of sources, sectors and gases would increase the number of available emissions reduction options and hence decrease the total costs for achieving a given emissions target.

Free allocation to the industrial sector would significantly decrease the total costs for compliance for this sector. Industry would be able to sell allowances to the transport sector and these revenues would be important. However, the impacts on industry of a higher allowance price would be unchanged, including higher marginal production costs, decreased output, altered investments and closure of installations.

However, even if we were to create a separate cap and trade system for transport and one for industry, as outlined in the preceding chapter, we can't completely isolate the two emissions trading systems from each other.

It is likely that the transport sector, indirectly, would influence the allowance price in the industrial ETS to some extent. The transport sector would need to find emissions reductions within its own sector in order to reach its emissions target. This may for instance include using bio energy for producing renewable transport fuels. Since bio energy is a finite resource, we can expect an increased price on bio energy. In the industrial ETS, bio energy is an important means for reducing carbon emissions, for instance in power production. The increased price of bio energy, caused by an increased demand from the transport sector, would clearly have an influence on the costs of emissions reductions in industry, which in turn would increase the allowance price. This effect is probably more significant for road transport than for aviation. We may also see an increased demand for transport-related electricity, for instance due to increased use of electric railways or plug-in hybrid cars. If this additional electricity were produced with bio energy, we could expect an even higher price on bio energy; and if it were produced by coal, it would increase the demand for allowances and consequently the price. A third example is that the transport sector may increase the global demand for CERs (Certified Emission Reductions), thus competing with industry. The CERs work as a common currency, linking different carbon regimes.

## 3. Summary

Greenhouse gas emissions from international aviation and shipping are growing at an alarming rate and little progress has been achieved in controlling these emissions. Meanwhile, substantial progress is being made in the development of international emissions trading. The EU emissions trading system is the main example, but we are also seeing other national and regional emission trading systems emerging, for example in Norway, Iceland, North Eastern USA, Australia and New Zealand. It is likely that there will be linkages between these systems. We may be seeing the first steps of what could become a global ETS.

With the slow progress of controlling international transport, it is natural to look at the possibility of linking international transport to a global emissions trading system. There are several advantages with this. Putting transport under a common cap would effectively put a limit to the total emissions and create incentives for emissions reductions in the international transport sector. Moreover, a broadening of an ETS to include more sectors is in general positive since it provides more options for reducing emissions.

But there are also risks for industry associated with such a solution and in this aspect we can learn from recent theoretical studies of integrating road transport into the EU ETS. The main factor for determining the economic consequences of linking two emissions trading systems is the relative difference in abatement costs. Even if there are large differences between EU road transport and international aviation and shipping, we believe that it is plausible that abatement costs or willingness to pay for transport are high in international transport relative to industry. If so, then the example of integrating road transport into the EU ETS holds many lessons for international shipping and aviation.

On the basis of this assumption, we consider that creating an integrated ETS with international transport and international industry, while potentially lowering the total costs for the included sectors to reach the common emissions target, could lead to considerably higher operational costs for industry and not necessarily reduce transport emissions. This may result in structural changes in the industry and possible carbon leakage even if this, of course, is reduced as we move towards a global ETS.

An alternative option is to apply a sectoral approach by creating a separate ETS for international transport alone. Thereby, transport emissions can be effectively controlled. The impact on the industrial allowance price can be reduced, at least to some extent. It is, however, likely that even with separate systems the transport sector will indirectly influence the allowance price in industry. Both sectors will compete for the same emissions reductions such as bio energy, clean electricity and CERs (certified emissions reductions). The transport sector is likely to increase the demand for these solutions and thus indirectly increase the allowance price in the industrial ETS. Another disadvantage is that the total costs for reaching a global emissions target will be increased as compared to an integrated ETS for transport and industry.

It is too early to judge which option is the best. Emissions from aviation and shipping need to be controlled and linking international transport to an ETS is an obvious option. This will be a crucial future challenge for the global climate change regime. But the issue is complex and there is no obvious best solution; therefore the issue will need further careful analysis.

## References

CE Delft (2006), *Dealing with transport emissions: An emission trading system for the transport sector, a viable solution?*, Swedish Environmental Protection Agency Report 5550, Stockholm.

Holmgren, K., M. Belhaj, J. Gode, E. Särnholm, L. Zetterberg and M. Åhman (2006), *Greenhouse Gas Emissions Trading for the Transport Sector*, IVL Report B-1703, Gothenburg: Swedish Environmental Research Institute.

Kågeson, P. (2008), *Transportsektorns koldioxidutsläpp och handeln med utsläppsrätter* (in Swedish), Forthcoming report to be published by Studieförbundet Näringsliv och Samhälle (www.sns.se).